Climate Change
Begins at Home

Life on the two-way street of global warming

Dave Reay

Macmillan

London   New York   Melbourne   Hong Kong

First published 2005 by
Macmillan
Houndmills, Basingstoke, Hampshire RG21 6XS and
175 Fifth Avenue, New York, N. Y. 10010
Companies and representatives throughout the world

ISBN-13: 978-1-4039-4578-5
ISBN-10: 1-4039-4578-0

This book is printed on paper suitable for recycling and made from fully managed and sustained forest sources.

A catalogue record for this book is available from the British Library.

A catalog record for this book is available from the Library of Congress.

10   9   8   7   6   5   4   3   2   1
14   13   12   11   10   09   08   07   06   05

Printed and bound in China

To Sarah

contents

acknowledgements

First among all, thank you Sarah and Maddy for the year's worth of weekends during which you put up with me shut in the bedroom, emerging only rarely for rocket-fuel coffee. Also, thank you Sarah for reading the draft chapters, for liking them and for laughing.

Thanks Flo for the head-clearing yomps over the fields. Thanks to my parents John and Joan for always being so interested, not to mention providing a great childhood with Mike, Paul and Elisabeth. Thanks to Glyn and Allan for all your support, good meals and great company. Thanks to Mahmoud Ibrahim, John McCrystal, Elspeth, Irene, Brian, Angie, Richard, Jenny, and all my good friends and neighbours in Threemiletown for making our little corner of West Lothian a Scottish Bedford Falls.

Thank you Keith Smith at Edinburgh University for being a friend as well as a boss, and for giving me the freedom to follow new ideas. Likewise, thank you to my other hugely supportive bosses at Edinburgh, John Moncrieff and John Grace.

Thank you to Scott, Mel, Kerry and Maud McEwan for all the afternoon walks and talks. Thanks also for, variously, help, advice and inspiration to: Simon Singh, Rachel Carson, Mark Lynas, George Monbiot, Richard Starkey, Paul Jacobson, Bernard Hutchinson, Douglas Adams, Gerald Durrell, the Intergovernmental Panel on Climate Change, the Natural Environment Research Council, and ABBA.

Finally, thank you to Sara Abdulla, my editor at Macmillan, for liking more than half of my jokes and for scraping the academic barnacles off my writing.

permissions

The author and publisher are grateful for permission to reproduce pictures from the following sources:

Intergovernmental Panel on Climate Change (IPCC). *Impacts of or risks from climate change, by reason for concern.* TS-12. Climate Change 2001: Impacts, Adaptation and Vulnerability. Published by Cambridge University Press, Cambridge, 2001. Reproduced with permission.

Paul (Jake) Jacobson. *Jacob Marley and Ebenezer Scrooge.* http://bcpub.com/jake/. Reproduced with permission.

Patrick Minnis, NASA. *Flight frequencies over the US on 3rd and 11th September 2001.* NASA Langley Cloud and Radiation Group, US. http://www-pm.larc.nasa.gov/sass/airtraffic_shutdown.html. Reproduced with permission.

Imperial War Museum. *Daddy, What did you do in the Great War?* (Saville Lumley). Poster held by Imperial War Museum, London. Reproduced with permission.

note on units

Throughout the book I refer to grams, kilograms or tonnes of greenhouse gas, by this I mean the amount of greenhouse gas as measured in so-called carbon dioxide equivalents. For this, emissions of powerful greenhouse gases like methane and nitrous oxide are multiplied by their global warming potential (GWP) which takes into account how much better they are than carbon dioxide at trapping heat on a molecule to molecule basis. So carbon dioxide has a GWP of 1, methane a GWP of 20 and nitrous oxide a GWP of 310.

A keen eye will also spot that while greenhouse emissions are given in metric, I have kept engine efficiencies in imperial units as, despite the best efforts of our governments, most people better understand fuel efficiency in terms of miles per gallon. Some other units are treated similarly.

preface

Climate change isn't new. Since the first microbes, drifting in a bubbling prehistoric ocean, noticed it was getting a bit chilly, life on Earth has had to either adapt to changes in the climate or die. What *is* new is a rapidly changing climate driven by an enhanced greenhouse effect – humankind's uncontrolled experiment on the planet whereby we pump enough greenhouse gas into the atmosphere to double or triple its concentration and then see what happens.

The reason I've written this book is simple: I don't want to see what happens. I don't want my family and friends to see it, I don't want you or your loved ones to see it. Most of all I don't want our children and grandchildren to see it. I'm frightened by what climate change has in store, really frightened.

I haven't always been this worried. For years my interest in global warming was more professional than personal. As a fresh-faced graduate my research centred on the descendants of those chilly microbes in the frigid waters of the Southern Ocean, asking: how will they respond to a warming world?

(Some liked it, some died.) For the next seven years, I carried on counting the climate change beans in the belief that politicians would see the need for action, and act. When the Kyoto Protocol emerged in 1997 I knew that, by itself, the treaty wouldn't be enough. Nevertheless, I felt reassured that with so many nations on board something would finally get done. Then, in 2001, US president George W. Bush withdrew his nation – the world's single biggest greenhouse gas emitter – from Kyoto. For a while I just wandered about at work muttering "What the hell happened?". At the time it looked as though Kyoto was dead in the water, that there was to be no concerted action on cutting global greenhouse emissions, and all the scientific research (my own algae boiling included) counted for nothing.

It would have been easy to continue feeding George W dolls to my Labrador and moaning about politicians, but there was a way to fight back. Alongside the research I was being paid to do, I started to look into the greenhouse gas that I was directly responsible for, the emissions which were mine to increase or decrease as I saw fit. I found that I was a big emitter, but also that I could do something about it. The politicians may have been dithering, but I was going to cut my own emissions – to do my bit.

And so were sown the seeds of this book. First came a short paper in the journal *Nature* called "Kyoto Begins at Home" about a family of four in the USA who met their own equivalent of the USA's Kyoto commitment through a few simple lifestyle changes. During the subsequent years I researched everything from green burials to the global warming contribution of Labradors. Along the way our big car was swapped for a Smart car, low-energy bulbs spread through the house, and the mail-order composting worms arrived (a fun evening in, I can tell you).

During this time I was also running a website covering climate change research. Each month the workload increased as the column inches that global warming was attracting shot up. These stories were no longer only about its impacts in the developing world. There were now headlines about Alaskan communities sinking into the ground, heatwaves killing tens of thousands across Europe, flash flooding in the USA, drought in Australia, and even bankrupt ski resorts here in Scotland. Climate change, it seemed, was coming home.

All this, along with some severe sleep deprivation, led to this book. *Climate Change Begins at Home* takes a long hard look at life on the two-way street of global warming and brings our impact on the climate face to face with the shifting climate's impact on us, our neighbours, and generations of first-time buyers to come.

1

the power of one

Meet the Carbones, a middle class family living in a typical street. A street, just like thousands of others in the Western world, with picket fences and trimmed lawns, with coffee mornings and a neighbourhood watch. The Carbones have worked hard to buy the house and cars they want, and to provide for their two young sons, George and Henry.

It's Saturday morning and the Carbones have just got back from the supermarket. They are unloading the week's shopping from their gleaming people carrier. The weather in Alabama is wiltingly hot again and it's a race to get the ice cream into the freezer before it melts completely. Once in the house, George racks up the air-conditioning and slides in front of the TV. Henry slopes off to his room, puts on some music and rejoins his online game. With shopping safely put away, and all the carrier bags stored for reuse, Mr and Mrs Carbone sit down with the newspapers and a fresh cup of fairly-traded coffee. Life is good.

Soon though, their chirpy conversation trails off and their buoyant mood is punctured by the doom-laden headlines. The *Greenville Herald* warns of the heatwave intensifying and prints an appeal from the local Senator for restraint in electricity use at peak times to avert a power blackout. Health officials estimate that 2,000 people across the county have already been hospitalised with heat stroke and 30 have died. They urge the public to keep an eye on elderly neighbours and not to leave dogs or young children unattended in cars. Two boys have drowned trying to cool off by swimming in a local lake. The national news isn't much better. Across the Mid-West there is strict water rationing, many farmers are watching their top-soil blowing away in the wind, and up in Alaska thousands of homes have been lost to thawing permafrost.

This haywire climate, not that of some distant African state, but here, at home, has the Carbones worried. Just five years ago both would have dismissed any suggestion of taking action on climate change, citing the economic arguments of politicians and hinting at the vested interests of climate scientists. Then, the warnings were dire but the evidence appeared scant – and both could still remember 1970s predictions of our falling into a new ice age.

Now, although the Carbones still feel that the threat has been exaggerated, they cannot deny the ever-earlier start to spring in Greenville, the disappointingly snow-free winters, and the scorching summer, which at that very moment is finishing off the last few living patches of Mr Carbone's formerly immaculate lawn.

As their worry about global warming has grown, the Carbones have become increasingly keen to 'do their bit' to prevent it. For a while now, Mrs Carbone has separated the bottles, tins and newspapers from the rest of their household trash and, each Thursday, filled the special crates for kerbside collection. Mr Carbone has replaced a couple of their traditional tungsten light bulbs with low-energy bulbs, and is always on at the boys for leaving lights, TVs and computers on. They may not wear knitted trousers and eat muesli for every meal, but the Carbones feel they do what they can for the environment and would describe their lifestyle as 'really quite green'.

So what help are the Carbones' various actions in mitigating climate change? The blunt answer is: not much. Their weekly car trips to the supermarket produce more greenhouse gas than all that saved by their efforts to recycle and cut energy wastage. The Carbones have fooled themselves into thinking that they can help tackle climate change without making real changes to their lifestyles. In fact, the only purpose their 'environmental' efforts currently serve is to massage their uneasiness about global warming while allowing them to carry on living just as they've always done.

Why don't they do more? Well, apart from finding some lifestyle changes hard to swallow, like getting rid of Mr Carbone's gas-guzzling SUV, they've tended to see climate change as a developing-world problem. Sure, they felt bad about the increase in hurricanes, droughts and floods in Africa, Asia and South America, but until global warming came knocking on their own door it was all too easy to ignore.

For the Carbones and millions like them, it takes a home visit to really stir up some action. Warn them that their greenhouse gas emissions may harm people thousands of kilometres away and they might try to do a bit – some recycling for instance – but then they can always avoid TV when the news gets too bad. Make them aware that climate change is likely to threaten their own friends, family and way of life though, and they'll be the first in line for low-energy light bulbs at the local hardware store.

This is the challenge faced by all those calling for action on greenhouse gas emissions. They can appeal to our humanity, listing the devastating consequences climate change will have, or is already having, in the developing world – try to tap that guilty streak that runs through all of us living in the luxury of the West. They can, and do, warn of the famine, plagues and migrations of biblical proportions that could occur as climate change takes hold in countries with scant or no resources to adapt. But as long as the apocalypses are remote many people will drag their heels.

Only when climate change starts to squeeze us directly will we really begin to take notice – when it's not only drought-stricken Sudan or flood-ravaged Bangladesh in the news, but our own neighbourhoods and economies taking a hammering. The most severe effects are still some years away and we have kidded ourselves into thinking we have ample time to head off any big problems. We haven't.

Like most families, the Carbones assume that unlobbied governments will deal with such global issues, or that scientists will come up with a technological fix – a silver bullet to solve climate change. Neither of these head-in-the-sand solutions is realistic. In the end the buck stops with you, me, and the Carbones: our lifestyles and our emissions. So how much greenhouse gas does the Carbone family emit and how?

Mrs Carbone

Kate Carbone's life is, to say the least, hectic. Now in her late 30s, she successfully combines bringing up two sons, the bulk of the housework, at least two-thirds of the dog walking and a job as team leader in a local travel firm. Most mornings, it's non-stop demands from the moment she wakes up, from the dog whining for its walk, through the muttering husband unable to find matching socks, to George and Henry arguing about who opens the cereal packet. At the centre of this whirlwind Kate directs proceedings and somehow everyone ends up where they should be with the minimum of forgotten lunches and inside-out sweaters. Running George to and from school is invariably the most stressful part of her day. He always waits until the last possible minute before dashing out to her eight-seater people carrier, despite any amount of horn honking, threats of being left behind, pocket money restrictions and bedroom incarceration. Once the family are all safely off, Kate makes her way through the rush-hour traffic to her office and, with a cup of strong coffee, sits down for another day of placating disgruntled customers whose shower had been too hot/cold, whose hotel was too far from the beach, or who had found it inconvenient that not all Germans are fluent in English. Evenings for Kate Carbone are only slightly less hectic, with after-school clubs meaning more ferrying about of children, dinners to be prepared and eaten, and battles to be fought over when homework should be done and how much TV can be watched.

At the weekend Kate spends as much time as she can in the garden – her pride and joy. Over the years she has transformed it from a featureless waste of brown grass and bramble-ridden borders into a riot of colour and buzzing insects. She has turned a section over to herbs and vegetables, providing the family with fresh salads for much of the summer. There are always a hundred and one jobs to get done, with progress never helped by bored sons, husbands and Labradors. But Kate, aided by regular visits from her mother-in-law, Grandma Carbone, somehow keeps the weeds under control and the flowers dead-headed.

Kate Carbone's climate impact is dominated by her car-driving. All those school runs, the slow drag into work and back, and the numerous food-shopping trips add up to over six tonnes of greenhouse gas each year. She does, though, cut her emissions by growing some of the family's food and so avoiding all the transport normally associated with it.

Mr Carbone

John Carbone is hugely proud of his family, his house and the life he has helped make for them all. It has been hard work, particularly in the early days when he and Kate had a young baby, only one insecure job between them and an income which could only just about cover the mortgage repayments. John still puts in long hours at the office and has his sights set on moving higher up the career ladder, but the money is now pretty good and he can take more time off without worrying that his job will be given to someone else. His day starts with a hasty shower and coffee to go, then it's into his brand new SUV and off to join the slowly moving queue of cars on the main road into the centre of town. John works for a large insurance firm and has recently been made acting manager of his branch. He reckons that in a year or so he should make full branch manager. Every year he flies to Seattle for the company's Annual General Meeting and, through some blatant networking, has earned himself the reputation as a rising star in the company.

At the weekends John usually has some paperwork to get out of the way. Once that's done its down to the usual weekend chores of watering the lawn, child ferrying, food shopping and DIY. He usually manages to squeeze in an hour or two of TV sport and, in the summer, a few beers with the neighbours and the odd barbecue.

Again, it's transport which dominates John's climate burden. His large-engined car clocks up a massive 12 tonnes of greenhouse gas emissions a year. His annual Seattle flight adds another tonne. At home he and his wife must take the blame for the bulk of greenhouse gas emissions, being responsible for how the house is heated or cooled, and for the efficiency of the many electrical appliances.

We all face the same problem as our families and/or our salaries grow. With greater affluence, the opportunity to increase greenhouse gas emissions through energy-rich activities also rises and the financial incentive for keeping down energy use fades. The climate impact of the two-kid, double-income Carbones has rocketed from those early days as newly-weds who always caught the bus to work and wouldn't dream of leaving a light on. Their household energy expenditure is now in the 'high-user' category, the resulting emissions adding up to 13 tonnes a year. With growing boys to feed; bought-in food alone totals over 30 kg of goods each week; the greenhouse gas arising from the transport of all this food adds up to about four and half tonnes a year. The household and garden waste that doesn't get recycled leads to yet another tonne of emissions as it slowly rots in the dark recesses of the local landfill.

George Carbone

George is nearly eight. What he lacks in age he more than makes up for in energy. He is the main reason the Carbones' front and back lawns are bedecked year-round with trampolines, slides, and all manner of other plastic tat. Play in the back garden can sometimes lead to clashes with his mother, particularly when a plant mysteriously loses its flowers heads or develops snapped stem disease. Being so young, George's climate contribution is mainly determined by his parents. His daily drive to and from school in Mom's people carrier clocks up over 600 kg of greenhouse a year. But this youngest member of the family is already damaging the planet through his waste of energy around the house. He frequently leaves on lights, TVs and countless battery-powered toys, despite the threats and pleas from his Dad. This squandered energy equates to an extra 120 kg of greenhouse gas emission each year.

Henry Carbone

Henry is 12 and becomes obsessed with any craze that passes through his high school. One week it is collecting game cards, the next skateboarding. A month later the oh-so-desperately-needed skateboard will be gathering dust in the Carbone shed while Henry clocks up his fourth straight hour playing an online fantasy game. His room looks like mission control; TV, stereo, PC, mobile phone and modem flash and hum away to themselves day and night.

Like his little brother, Henry routinely leaves his gadgets on, whether in use or not, resulting in 160 kg of greenhouse emissions each year. He's also got into the habit of turning on the electric radiator in his room at the first hint of chill through his Black Beelzebub T-shirt. This additional two hours of heating each day pumps out 700 kg of greenhouse gas every year.

Fortunately, Henry's decision to travel to school by bus (it gives him more time to swap Orc Ascension cards) saves more than half a tonne of greenhouse gas compared to going in his mother's car. The bus emits only 53 kg to get each card-sharp to and from school for a year.

Molly

Molly is the Carbone Labrador. At nine years old her penchant for socks means no one in the family can claim to have more than a couple of matching pairs (and these all have holes in). Molly's climate impact is wholly the responsibility of John and Kate Carbone. It centres around whether or not they choose to drive to her favourite walking spots. They usually do, as most of the best are either too far or too dangerous to walk to.

An average walk entails a 6 km drive in the people carrier (the SUV is too new for muddy Labradors), which belches 300 g of greenhouse gas for every kilometre it travels – about 4 kg on each round trip. With two such walks every day of the year, rain or shine, Molly's annual emissions soar to nearly three tonnes.

Grandma Carbone

Grandma Carbone lives about half an hour's drive south from Kate, John and the boys. She has dwelt in the same large, rambling house for over forty years and every corner is home to a happy memory. Grandpa Carbone built a good deal of it with his own hands, back when most of the land about was still fields and the traffic-clogged road at the front was just a dirt track. As a young couple, Grandma and Grandpa Carbone worked like Trojans to keep up the loan repayments. They spent all of their weekends painting, decorating and planning more improvements. Their son, John, was born and raised in the house, and he and Kate had their wedding reception in its garden. Since Grandpa Carbone's sudden death eight years ago, Grandma Carbone has lived alone in the house – not that she spends much time there, being out almost every day at some fundraiser, playing golf, or helping out with the grandchildren.

When Grandpa Carbone died, Grandma swapped his pride and joy – a 1968 gas-guzzling classic car – for a bright yellow hatchback in which she clocks up around 9,000 kilometres a year. This new car emits two tonnes of greenhouse gas a year, a great improvement on the old one which drank twice as much fuel and was a pig to park in town.

Grandma Carbone flies each Easter to visit her sister in Beaverton, Oregon, adding about a tonne to her own annual greenhouse gas budget in the process. Since the death of Grandpa Carbone, energy use at home has dropped into the 'low' household bracket with the related emissions adding up to five tonnes a year. The old place is feeling too big and Grandma Carbone has plans to move into a more up-to-date and compact retirement flat.

Though her son and daughter in-law have mentioned things like recycling and 'the environment' on numerous occasions, Grandma Carbone has no truck with such lefty concepts. She does hanker for the cooler summer days of her youth, but not the enforced thriftiness of the war years. As such, her green-house gas emissions add up to ten tonnes each year and the recycling boxes in her backyard collect only dead leaves and rain water.

Over at Kate and John's house where, unlike Grandma, they are 'doing their bit' for the environment, how do they com-pare climate-wise? Let's take the good things first. They recycle most of their newspapers and cardboard, saving about 400 kg of greenhouse gas per year. The recycling of their glass bottles and jars each year, along with all the tin cans, cuts another 300 kg off their annual greenhouse budget. Kate Carbone grows enough salad and root vegetables in her garden to prevent 300 kg of greenhouse gas that would have been emitted from production and transport of this food. John has fitted three energy-efficient bulbs and intends to get around to replacing them all, so far cutting 225 kg of greenhouse gas off the yearly family budget and bringing their total reductions to just over 1,200 kg.

On the down side, the family's emissions are dominated by transport: between them their two large-engined cars pump out a total of 18 tonnes of greenhouse gas per year. And then there is the summer holiday. For the past six years, the entire Carbone clan, (apart from Molly who goes into kennels), has jetted off to Cancun in Mexico for two weeks' swimming and sunbathing at the same self-catered house. For every kilometre flown, each family member racks up a further 150 g of green-house emissions. So the round trip to Cancun adds 400 kg to each of their annual budgets. Throw in John's AGM jaunt to Seattle and family air travel produces more than two and a half tonnes of greenhouse gas a year.

Totting things up, we have 20.5 tonnes of greenhouse gas from transport, 13 tonnes from home energy use, four and a half tonnes of food-related emissions, and the tonne produced due to household waste. In other words the 'we're doing our bit' Carbone household puts 39 tonnes of greenhouse gas into the atmosphere each year – enough to fill their house 40 times over.

The Carbones have managed to cut their greenhouse gas emissions by an embarrassingly small 3%. This doesn't even get them to the 5.2% target that politicians have set with the Kyoto Protocol, let alone the 60% cut scientists urge. It's a shock to a family who felt that they were doing well in mitigat-ing their damage to the environment. The lifestyle changes they actually need to make go a tad beyond leaving the news-papers out for collection once a fortnight.

↔

The south-eastern USA, home to the Carbones, has been a boom area in the past few decades. It produces about a quarter of all US-grown food and half of the timber. The population has shot up by around a third since the 1970s, with most people moving into new homes along the coast. Climate change poses a real threat to hundreds of thousands, maybe millions, of people in this region, endangering their homes, their jobs, and even their lives.

Summertime temperatures across the south-eastern states are projected to soar by more than 10 °C by 2100, the most dramatic rises predicted anywhere in the USA. Such big hikes threaten people right across the region, especially the young, old and poor. As temperatures soar, so does the danger to those without expensive air-conditioning systems and those physically more vulnerable to heat stress. In Atlanta, Georgia, for example, temperatures on a sunny July day could top 54 °C during this century, unbearable for the most sturdy lumber-jack, let alone a new-born baby or its bedridden grandpa.

The huge energy demand from those who are lucky enough to have air-conditioning will mean more air pollution from power stations, and therefore more respiratory health problems due to low-level ozone and particulates. All in all, not the ideal recipe for a carefree summer in the south-eastern USA.

Another big concern is flooding. It is already a leading cause of deaths due to natural disasters in the south-eastern USA, and in the USA as a whole, causing about 100 fatalities each year. Sea level has risen almost 30 cm since the mid-19th century and models suggest it could rise a further 90 cm or more by 2100. Around 400,000 hectares of wetland have disappeared in the past century. Many hundreds of hectares of coastal forest have been destroyed, the saltwater slowly intruding inland, killing off the trees as it goes.

In recent decades, a combination of rising tides and land development has already led to the loss of 13,000 hectares of coastal salt marsh – the natural buffer that protects the land from flooding and erosion by the sea. As sea levels go up, yet more of these natural barriers will be lost, increasing the flooding threat to communities further inland. Rainfall is also expected to increase by about 25% by the end of this century, further adding to flooding risks.

Water quality right across the region is imperilled by climate change. Salt water is increasingly likely to contaminate drinking water supplies near coasts, while higher temperatures reduce the amount of oxygen in the water, damaging fish and other aquatic life. Flash flooding can lead to pollution of drinking water supplies by sewage, rotting carcasses, chemicals and fuel. North Carolina suffered exactly this fate in 1999.

The economic costs of such climate change sequelae are set to grow and grow. The bill for weather-related disasters in the south-eastern USA has topped $85 billion over the last 20 years – the heatwave and drought of 1998 alone cost $6 billion and 200 lives.

Crop yields are expected to rise in some areas but fall in others, particularly the Gulf coast, during this century. Some soybean farmers have already seen harvests drop 80%, while those growing irrigated wheat may see rises of 20% by 2090. Higher concentrations of carbon dioxide in the atmosphere are likely to boost tree growth in the south-east; models predict increases of up to 10% in pine production and up to 25% in hardwood production in the next 100 years. But there's bad news too for our sweaty lumberjack. On top of the destruction of forests by salt water, drier soils and more fires (brought about by those soaring summertime temperatures) will probably see large areas of forest replaced by grasslands by 2100.

The Carbone house is 160 kilometres from the coast and, at about 130 m, well above sea level. But, with heavier rainstorms predicted, the local river poses a serious flooding risk. Through his work in insurance John Carbone is all too aware of the danger. Since the last scientific assessment of the flooding risk in Alabama was published, his own insurance premiums have rocketed and almost every day he has to tell a potential customer that, as their house is within 30 kilometres of the coast, it is effectively uninsurable.

Grandma Carbone has suffered water contamination half a dozen times in the past 10 years, each time having to rely on rapidly diminishing supplies of bottled water from the local supermarket. John had to upgrade the air-conditioning system throughout his house last summer; the existing system couldn't cope with being run at full blast for such long periods and eventually gave up the fight completely – leading to a very sticky couple of days trying to persuade the air-conditioning company to fit a new one when they had hundreds of similar requests from across the county.

Soaring summer temperatures will also affect the Carbones' holiday. Though models of the effects of climate change are best for developed-world countries, it is clear that Mexico will be hit hard. Summer heatwaves are expected to intensify and cause many more deaths, mainly through heart and respiratory illnesses. The problem is exacerbated in Mexico by the general lack of air-conditioning, high population density and a fragile health-care system. Things will likely be worst in Mexico City and other large, overcrowded, polluted urban areas. Cities experience the 'heat island' effect: roads and buildings absorb and trap warmth, keeping temperatures several degrees higher than in the surrounding countryside, even at night. This, combined with overall temperature increases, is likely to push up average temperatures in Mexico City by 5 °C during this century. Add in the very real threat to water supplies across Mexico, as increasing demand clashes with falling availability, and you have the recipe for a human health disaster.

Around 18 million US citizens currently holiday in Mexico every year. Like the Carbones, these tourists will start staying away in droves if the summer temperatures rise too high. This is the last thing Mexico's fragile economy can afford, given the huge costs it faces in coping with climate change.

The threats facing the Carbones and their neighbours in the south-eastern USA will be echoed around the world. For some, things will be much worse, for others, not as bad. The precise impacts of global warming are often hard to predict because the world's climate system is so complicated. One effect may cause another, resulting in a cascade of problems. Some effects may cancel out others. The global evapo-transpiration cycle is expected to speed up. This means that generally it will rain more, but water will evaporate faster – leaving soils drier during critical parts of the growing season and potentially causing widespread crop loss and famine.

Maximum wind speeds will increase and extreme weather events, such as hurricanes, may occur more frequently. Rising sea levels, from thermal expansion and/or ice-shelf disintegration, may result in mass migrations and the loss of huge tracts of farmland. Some islands and low-lying coastal regions may disappear altogether: Tuvalu in the South Pacific and the Sundarbans delta in Bangladesh, for instance. Global average sea level has already risen by around 15 cm over the past 100 years, with another 18 cm rise predicted due to global warming in the next 30 years. If this trend continues, we could see a rise of up to 88 cm by 2100.

It's the poorest countries of the world – those least responsible for human-made climate change – that are likely to feel the most severe effects. Many areas will lose huge swathes of agricultural land to flooding, freak weather and pests, leading to starvation on an unprecedented scale. Clean water will become ever more scarce, which, in combination with poor sanitation, over-stretched medical services and bad nutrition, will provide a perfect environment for disease epidemics.

Expanding deserts, extreme storms, floods and starvation could displace many millions of people, causing civil unrest and a rising pressure on immigration systems. Tropical diseases are expected to spread to new areas. The geographical zone of potential malaria transmission, for example, today covers around 45% of the world population; by 2050 that figure may be 60%. Finally, climate change will have a massive impact on ecosystems and wildlife right around the world, with the climate-warming scenarios for 2050 indicating that up to a third of all land species will by then be committed to extinction.

Long before the earliest hairy-backed hominids started lighting fires, our atmosphere already had greenhouse gases and a fully functioning 'greenhouse effect'. Just as the glass of a greenhouse lets the sunlight through but keeps a good deal of the reflected heat from escaping back outside, so our atmosphere lets the light of the Sun through but catches some of the thermal radiation (heat) emitted back towards space from the Earth's surface.

Look out of the window. If you're lucky the Sun is shining down and the birds are singing. Everything you can see, from your driveway and your patchy lawn, to the neighbours' car and their novelty garden toadstools, is throwing energy back towards space. The amount of energy bounced back to the atmosphere in this way has to balance the amount coming in from the Sun; if less is able to leave than comes in, as happens when greenhouse gases trap some of it, the Earth starts to heat up – global warming.

Without this warming the average temperature on Earth would be a frigid −18 °C rather than our more balmy average of 15 °C. The big problem though is that the greenhouse gas blanket that has been keeping the Earth warm for millennia is

Figure 1 The Monkey Puzzle Tree (*Araucaria araucaria*). Busy doing its bit to lock up carbon dioxide.

being added to on a massive scale by humans. Take a look at the historical records of the amounts of the various greenhouse gases in our atmosphere and all the graphs have the same shape. They bumble along for thousands of years and then, around the end of the 18th and beginning of the 19th centuries, they start to rise, slowly at first and then faster, like a stock price in a world of boom and no bust. With the Industrial Revolution came a huge growth in the burning of fossil fuels. The amounts of carbon dioxide have risen by almost 30%, while methane – a greenhouse gas produced by rice-growing, landfills and burping cows – has leapt to more than double pre-1800 concentrations. The quantities of these greenhouse gases in our atmosphere are set to carry on increasing long after you and I have shuffled off this mortal coil. Carbon dioxide concentrations, for instance, are expected to more than double by 2100. As a result, global temperatures will rise by between 2 and 5 °C, having already gone up by about half a degree during the 20th century. It's this warming, and the catastrophic impact it may have on our global climate, that threatens us all.

Let's take a closer look at the key players in this – the greenhouse gases themselves. Carbon dioxide and methane are both big hitters, but there are another four that are also of real concern globally, and which politicians have specified as in need of control. First is nitrous oxide, commonly known as laughing gas. Much of it comes from fertilisers and cattle feeds. Then come the catchily named hydrofluorocarbons (HFCs), perfluorocarbons (PFCs) and something called sulphur hexafluoride (SF_6). These last three are entirely artificial and come from things like fridge coolants and propellants. Ironically the HFCs were introduced to replace the ozone-damaging CFCs that once chilled fridges and fired out our hairspray. Together these gases make up the so-called 'basket of six', each of them threatening our climate and all of them produced in part or wholly by us. (Water vapour is another powerful greenhouse gas, but, as our effect on it is largely via emissions of the other gases, it is left out of the greenhouse gas 'basket'.)

Some greenhouse gases are much better at trapping the heat radiated from the Earth than others. The most famous, carbon dioxide, is not actually the best heat-catcher but is present at higher concentrations than the others and so plays a bigger role in global warming. The HFCs on the other hand occur in vanishingly small amounts, but each molecule can be thousands of times more effective at trapping the heat radiated by the Earth than a molecule of carbon dioxide.

Take a look into the sky. Imagine all that radiated energy streaming back towards space, invisible to our eyes but on its way nonetheless. Up it goes through our atmosphere, much of it back into the cold vacuum of space, but some – that encountering a greenhouse gas – has a different journey. Think of what happens as like energy trawling, where the energy emitted from the Earth is like a shoal of fish making a dash for the cool of space. All those greenhouse gases in the atmosphere are like giant nets spread across the sky. The carbon dioxide net, though one of the biggest, has a rather coarse mesh and so lets most fish through. Methane, nitrous oxide and the others, despite having much smaller nets, also have a much finer mesh and so are much better at trapping our, by now, rather alti-tude-sick fish. Methane's net for instance is effectively 20 times finer than that of carbon dioxide, while the silk-stocking-esque nitrous oxide net is 300 times finer. To stay with our fishy example, humankind's emissions of carbon dioxide and the other greenhouse gases over the last century or so have given those atmospheric trawlers the equivalent of bigger nets, finer meshes and a catch which is growing by the day.

As emissions and global temperatures increased, and the pre-dictions worsened, the international community, under the umbrella of the United Nations, began to wake up. In 1988 the Intergovernmental Panel on Climate Change (IPCC) was estab-lished to assess the risk of human-induced climate change. The UN Framework Convention on Climate Change (UNFCCC) was developed to provide the overall policy for addressing cli-mate change and so forms the foundation of global efforts to combat global warming, such as those set out in the Kyoto Protocol.

The Kyoto Protocol was drawn up in 1997 by more than 160 nations to set targets for global reductions in greenhouse gas emissions.

Targets in the Kyoto Protocol ranged from a 7% cut for the USA to a 10% increase for Iceland. The global target was for a 5.2% emissions cut, but the scientific consensus is that a world-wide reduction of 60% is required by 2050 to stabilise our climate and head off catastrophic impacts. Achieving such a cut could, for example, reduce the amount of flooding along the Bangladesh coastline of 2100 by 90%, compared to taking no action. The 60% cut is undeniably big. If we're going to achieve it we need meaningful action on emissions right now.

The Kyoto Protocol cuts are a start, but even they are being undermined. Two of the world's biggest emitters, the USA and Australia, have refused to sign up even to the small targets of Kyoto, while loophole exploitation by countries who have signed threatens to reduce the value of cuts set out in the Protocol even further.

To recap: thousands of top-notch scientists from all over the world are warning that if we don't reduce greenhouse gas emissions, we, our children, and our children's children, will almost certainly suffer dire consequences. Getting a wide array of science's big guns to agree on anything is like herding cats, yet on climate change they have reached this sobering consensus:

most of the warming observed over the last 50 years is likely to be attributable to human activities.

This strong statement from an inherently conservative group was made back in 2001. And still emissions increase.

In the interests of even-handedness, I should mention that there are still some scientists who believe we shouldn't worry about global warming. There are about six of them. This small yet vociferous clique argues that our climate has always been changing and that the warming we see is simply a result of the natural fluctuations in global temperature. This sounds plausible, but their argument collapses though when you look closely at just how fast the Earth is warming. Include all the 'natural' things that control the planet's temperature – things like solar activity, volcanoes, and how close we are to the Sun – and you're still left with a big missing source of heating. The culprit? Artificially generated greenhouse emissions.

The climate change nay-sayers can babble on about natural variations all they want, but here are the bare facts:

- Greenhouse gases warm the planet.
- Global temperatures have risen 0.6 °C in the last 100 years.
- Concentrations of greenhouse gas in our atmosphere are now higher than at any time in the last 420,000 years.
- Since the Industrial Revolution greenhouse gas concentrations have risen by around 50%.

'Natural' variations? They've got to be kidding.

There may only be a dwindling handful of researchers who continue to deny the existence of human-induced climate change, but they have some powerful political allies. The US administration has often cited the 'scientific uncertainty' in defending its anti-Kyoto Protocol stance. For them the 420,000 year record is too short, still too uncertain. As I write, colleagues across Europe are working on an ice core from Antarctica that contains an even longer record of greenhouse gas concentrations in our atmosphere – almost a million years of climate history. The sceptics want more proof? They're going to get it.

Developed-world nations (essentially the industrialised West) are home to only about 20% of the world's population, but use about 80% of the world's resources. Ominously, the oil-powered standard of living we enjoy is what many in developing countries strive for. If the billions of people in the developing world progress to the same level of consumption and emissions as an average US citizen, we will be looking at climate meltdown. China, a nation of over 1 billion souls and possessor of huge coal reserves, is already the world's second largest emitter of greenhouse gas and is gaining fast on the USA.

That the warnings from the scientific community have not yet elicited a meaningful global response to climate change is astonishing. The Kyoto Protocol lives on, but still without the world's largest emitter on board. All is not lost though. A large, and growing, number of people around the world are starting to take their own steps to tackle global warming. The actions of individuals on a global scale could have a profound effect. Domestic and private transport emissions are huge, so individual action can have a big impact. Add to this the bottom-up effect that such lifestyle changes can have on community, business, and eventually government climate policy, and the huge importance of individual action on climate change is clear.

Throughout much of the past three centuries, we in the developed world have been emitting ever-increasing amounts of carbon dioxide, methane and nitrous oxide into the atmosphere, with no comprehension of the potentially disastrous consequences. We no longer have the defence of ignorance. The pressure is now on to turn off that light, get out that bike, and really 'do our bit'.

Why should you take action? Because it's you who will suffer the consequences if you don't – your family, your friends and most definitely your descendants. Our parents' and grandparents' generations will be remembered for huge technological changes, for great leaps forward in agriculture and medicine, and for fighting two world wars that allowed many millions of us to grow up in freedom. We, on the other hand, will be remembered as the 'selfish generations'. The ones who knew the damage their fossil fuel-driven lifestyles would do to the future but who carried on regardless. Our children will be angry with us, our grandchildren more so. And they will have fair reason. It brings to mind the War Office poster (Figure 2), shaming men into enlisting in the First World War. When your granddaughter asks "Grandad, did YOU drive a big car?" your face could well look like his.

Some time soon – maybe it's already happened somewhere – as a horde of children slope into their classroom and take their seats for another indigestible helping of double history and the Industrial Revolution, a teacher is going to surprise them. Instead of launching into a lesson on the significance of Jethro Tull and Thomas Edison to the food we eat and the lights we read by, the teacher will talk about another, very modern consequence of our ancestors' enlightenment – climate change. That class would go something like this:

Early in the 18th century night held sway over working practices and lifestyles. The setting of the Sun was usually the signal to cease work, with most jobs becoming impossible due to the lack of good-quality light. By 1721 the first 'factory' was in operation, signalling the start of the industrial era. By the late 18th century technological innovations, including the steam engine and the Spinning Jenny, were transforming production methods and working practices. Later, in 1831, Michael Faraday made generation of electricity practical and defying the darkness of night became possible on a large scale. Allied to changes in working practices and lifestyles during the 18th century

Figure 2 Daddy, what did **YOU** do in the Great War?

was a large increase in both national and individual greenhouse gas emissions. With the spread of industrialisation across Europe and North America, rapid human-made global warming was born.

Along with rapid population increase in the 18th century came a big rise in consumerism. Indeed, the most important driver of the late 18th century iron-smelting boom was not industrial demand for machine parts, ship hulls and the like; rather, it was consumer demand for domestic hardware – pots, pans and fireplaces. More iron meant burning more coal and so rising carbon dioxide emissions.

The new generation of steam engines in the 18th century were coal-powered themselves and pumped mineshafts dry to allow yet more coal extraction. In many ways, coal both fuelled the Industrial Revolution and paved the way for the global warming we are now experiencing.

Per capita greenhouse gas emissions rose rapidly during the 18th century. In England, for example, individual emissions rose from around 1 tonne per person per year in 1700, to around 3 tonnes per year by 1800. In the 19th and 20th centuries our rate of development and emissions grew hand in hand. The advent of electric lighting, motor cars, refrigerators and the rest marked new highs on this inexorable climb. From the base camp of 18th century steam engines to the gadget-littered summit of today, it's a climb that's become steeper by the year. In the UK, per capita greenhouse gas emissions now stand at around 11 tonnes each year; in the USA they top 20 tonnes.

So, what makes up an individual's climate impact? As we saw with the Carbones, for most of us it centres on our behaviour at home and our burning of fossil fuels to get around (Figure 3).

Topping the lifestyle chart at close to half of all our greenhouse gas emissions is transport. The chief culprit here is the car: all those gas-guzzling saloons, people carriers and four-wheel drives make up a major part of the climate burden for the average developed-world family. Indeed, for those of us with big cars our love of driving can constitute well over half of all our contribution to global warming. Our deepening affair with flying, for work and pleasure, has also become a significant part of most people's emission budgets.

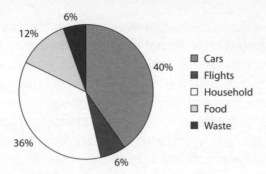

Figure 3 What makes up your own climate impact?

Energy use at home is the next big hitter, accounting for over a third of emissions. Nearly half of this is due to the heating and cooling of our homes. Refrigerators, freezers and all the other appliances and gadgets come a close second. These appliance-related emissions are increasing, as we fill our kitchens with ice-cream makers and giant refrigerators, our living rooms with huge plasma TV screens and our shelves with music centres and computers.

Next comes water heating at about 15% of household emissions, depending on the fuel used. Gas, for instance leads to far less carbon dioxide than does electricity from a coal-fired power station. Lighting comes in at 5–10% – more for those with a penchant for floodlighting their property. Cooking and clothes washing complete the picture.

Food makes up another large slice of our total greenhouse gas pie. The expectation of year-round availability of everything from kumquats to lemon grass, via salmon, tiger prawns and wild boar, means that the items of food in an average shopping basket can collectively have travelled 240,000 km, giving them a huge climate change tag. Add to this the methane emissions from belching cows and the miasma of nitrous oxide rising from fertiliser-soaked farmland, and food is responsible for 10–20% of our impact on the climate.

Waste is responsible for between 5% and 10% of our greenhouse gas emissions. Primarily this is because old food and newspapers transported to landfill sites rot down to produce the powerful greenhouse gas methane. But every shampoo bottle, drinks can and toothpick that goes into our bins also took energy to produce it, and so represents even more greenhouse gas.

So, that's where it all comes from. Between 10 and 20 tonnes of carbon dioxide, methane and nitrous oxide a year for most of us, and a good deal more for jet-setting, SUV-driving types. Can we really make big cuts in these emissions? Sixty per cent cuts? Is it expensive? Are electric cars the answer? Is recycling a con? Let's find out....

2

going places

Car and air travel, as with the Carbones, top most people's greenhouse gas budgets and so represent a huge opportunity for individuals to make a real difference. Unfortunately our way of life now so revolves around cars and planes that changing habits really does mean a break from the norm – taking the path less travelled. Look at the roads leading to your local mall: there are no pavements. Even if you wanted to walk to the shops you couldn't – not without risking being mown down by an SUV. The same is true of cycling. It's positively suicidal to ride a bike on many roads (Figure 4). Unless you are very lucky (or Swedish) there will be just a few kilometres of cycle path in your local town, probably scattered with broken glass, parked cars or potholes, or all three. Towns, cities, shops and offices, and even homes, are increasingly designed to fit the needs of that four-wheeled wonder that is the car, at the expense of less climate-damaging ways of getting around.

We've heard the pleas from politicians to use public transport, reduce speed, and drive a smaller car so many times we've become almost deaf to them. To date the effects of climate change on transport have tended to take the form of having the in-car air-conditioning on non-stop in the summer. Those of us who rely on an open window as opposed to a

Figure 4 The joys of cycling.

whirring piece of cooling technology under the bonnet have certainly been feeling the heat in recent summers. Higher temperatures, though, are just the start of a series of impacts on our car driving.

Those journeys through torrential rain, gripping the steering wheel for all you are worth as another wall of water from the truck in front obliterates your view – expect more of them. Likewise, there will be more weather-related accidents and 'road flooded' notices. And climate change will attack the very road surface itself. If you don't find yourself in a jam waiting for a fallen tree to be cleared you're likely to end up in one caused by repairs as the surface melts in summer, cracks in winter, or sags and disintegrates due to both. Bad news then if you think of driving as an efficient, safe and rapid way to get around. It's also going to get very expensive. As oil runs out its price will go up; road tax will climb to pay for all those repairs, insurance premiums will rise to cover increased accidents and, if governments are brave enough, drivers will face escalating taxes on their greenhouse gas.

Relying on top-down government action to cut our emissions from transport, though, could mean a very long wait. Yes, private road transport, being mainly cars, is a major source of greenhouse gas globally, but curbing our driving addiction has proved a political hot potato. There are more cars in the USA than people to drive them: the average household has 1.8 drivers and 1.9 personal vehicles. That's over 200 million private vehicles in the USA alone. Lined up end to end these would stretch around the Earth more than 20 times. Over a fifth of them are SUVs, and another fifth are trucks, meaning big engines and bigger emissions. The average US car does about 20 miles per gallon – the worst 4×4s do only 4 miles to the gallon – and, altogether, transportation in the USA creates nearly 2 billion tonnes of greenhouse gas each year. This is more than the total emissions of any other country, except

China (a nation with four times as many people). Each day, on average, an American will make four car trips totalling about 65 kilometres. Together the nation's drivers clock up about 17.5 billion kilometres each day.

The trend is repeated around the world. England's leafy lanes are now more likely host to four-wheel drives and people carriers than nippy Minis driven by Michael Caine lookalikes. The bustling streets of Paris, once the honking-ground of compact Renaults and Citroëns, are now patrolled by the same hulking four-wheel drives you'll find on a Texan highway. The sleeping greenhouse gas giant of Asia is waking up to aspirations of mass car ownership and US lifestyles. This is very bad: with more than 1 billion people in India and another 1.3 billion in China, if they all drove cars too....

Talk about how bad cars are for the planet and most drivers will respond that their public transport services are either poor or non-existent and that government first needs to provide viable alternatives. But, just as the matter is too urgent to sit and wait for politicians to take decisive action on climate change as a whole, it's also too pressing to wait until they've put a shiny new bus stop outside our house and provided a reliable service to the office every half an hour. So, what to do? I could launch into berating car users for their damaging and wasteful ways, demanding we change to a fossil-fuel-free society, where our children would be healthier and safer, and one could cycle to work without needing full body armour. When you look at the figures – the amounts of greenhouse gas we are emitting and the predicted consequences – it's hard not to take this kind of radical line. Let's face it though: if all I did here was tell you that you shouldn't drive, you shouldn't fly and that foreign holidays were out of the question, you'd fleetingly think about what this would mean to your own life, shove the book down the side of your bed, and get out the 'Learn Esperanto While You Sleep' tapes.

Cutting emissions, as with all addictive habits, takes awareness and willpower. You're a smoker and you like smoking, but
you also know that if you carry on smoking it will probably kill
you, so you'd like to give up. Going overnight from a twenty-a-
day user to a gibbering cold turkey doesn't appeal to most
people, and often doesn't work in the long run. Instead, you
cut down more gradually, you chew gum, put on patches,
avoid the pub, whatever makes the transition easier. Then one
day you get to a stage where the first thing you think of when
you wake up isn't a cigarette, and you know you can do it.

Some lifestyle changes may seem so small as to be irrelevant,
and on their own they are, but they represent that first nicotine
patch – a start. Right now we live in a car-dominated culture
with very limited public transport. It's very difficult for most of
us to simply forgo cars altogether and a fantasy to believe that,
like the cold turkey smoker, we can all just kick our car habit
overnight. Try telling a harassed mother of three, lugging her
shopping bags and children a mile home from the nearest bus
stop through driving rain, that it's OK because she's doing her
bit for the planet.

Reliable, affordable and accessible public transport is vital in
tackling climate change and reducing traffic congestion. But in
the USA, for example, public transport accounts for only 1% of
the billions of kilometres travelled each day. In Canada, the UK
and Australia the picture is similar, with public transport falter-
ing and car use growing.

Each time a bus or train route is abandoned, it makes it
harder for those without a car to get to work, the shops and
school. We live busy lives, and it is hard to relinquish the sheer
convenience of being able to get in the car, rather than trudg-
ing down to the bus stop to stand in the wind-blasted and ill-
named shelter for twenty minutes looking at a poster advertising the latest BMW and trying to avoid the gaze of the rather
worrying man with a ferret on a lead. Eventually, benighted

public transport users give in and get themselves a car, so the bus or train company loses yet more customers and their ticket money, leaving them less cash to run services; so they make further cutbacks. This ever-diminishing circle culminates in sparse networks of deserted (apart from the ferret man) or crammed-full buses and trains, and roads jammed with honking cars.

To break this cycle takes top-down (government) and bottom-up (you and me) action. The government must subsidise public transport, ensuring a good network that gets us where we want when we want; and you and I must use it. George Carbone's decision to take the school bus, rather than demand a lift in his mother's car, slashed his contribution to global warming. The same applies to us all. If you are lucky enough to still have a usable bus, train or even tram route to work then you can cut tonnes off your annual emissions. The exact savings vary depending on the service. But if you can use public transport, do.

Take a standard daily commute into work, say a 30 kilometre round trip. Your shiny saloon is in the drive, *ABBA's Greatest Hits* is teed-up in the CD player and it looks like rain. Driving is tempting, but the roads are awful, parking a nightmare and the train station is just five minutes' walk. By grabbing your umbrella and strolling down to catch the train rather than using the car you're cutting over 7 kg from your greenhouse emissions each day. Keep this up, and over a working year you've stopped a massive tonne and a half of greenhouse gas from getting into the atmosphere, helped keep a good train service running and may even have broken that worrying ABBA habit.

The effects of such changes add up fast. In 2003 London introduced congestion charging to discourage people from driving into the capital and encourage use of public transport. This pushed around 29,000 extra people out of their cars and onto the capital's public transport, mainly the beefed-up bus network. Transport-related greenhouse gas emissions in the charging zone fell by around 20%, the city became a nicer place to live and work and scores of other cities realised that you could actually take on the seemingly omnipotent car and win.

Along with congestion charging there are a host of ways in which the authorities can help, rather than hinder, our efforts to cut greenhouse gas emissions. For their part, the planning offices around the world could make sure that there are foot-paths when new buildings go up and that cycle lanes become standard in towns and cities. Instead of architects spending so much time showing how the 10 acre car park of every new gleaming office block could be 'softened' with a few plants they could be made to provide details on how public transport will be used to get there.

As long as trains are expensive and late, and buses infrequent, uncomfortable and inconvenient, we will keep picking up those keys, starting up our self-contained transport bubbles and bursting into another rendition of 'Dancing Queen'. The day when the hard-line environmentalists can be justifiably appalled at any use of cars is around the corner, but at the moment many of us have at least a passable defence in the shocking provision of alternatives.

So, for now at least, you plan to go on using your four-wheeled cocoon. In it you want your chosen music, your per-fect temperature and to breathe your own recycled air. OK, but you can have all this without an engine big enough to power a tank and that spews out enough greenhouse gas each year to fill the office block you're driving to. Increasing the efficiency of the average car from around 20 miles per gallon to a not unreasonable 40 mpg would cut US emissions by 500 million tonnes. This is without altering the number of cars, or the comfort of their drivers – assuming they can bear not having to take up two parking spaces wherever they go and put up with climbing into their cars without the use of ladders.

John Carbone is facing up to this. His four-wheel drive is, or rather was, his pride and joy. The car cost nearly a year's salary and came with all the added extras. But in recent weeks the pleasure of washing it each Sunday, polishing its chrome-work and cruising to work and back has begun to diminish as his worry about climate change has grown. Now, each time he reaches down to flick on the 'Climate Control' button he shifts uncomfortably in his seat and turns up the radio. For John, the car is turning from a badge of status and power to one of ignorance and selfishness. For years he had been the first around the water cooler to bring up the subject of cars and moan about the price of gasoline. In the months leading up to buying his SUV he became an expert on different engine sizes, top speeds and even colours available, and imagined a thousand times the feeling of sitting high up in the new car, the engine roaring at the turn of the key and Eric Clapton on the radio. On every drive he had kept a special eye out for other models, admiring their relative merits. Now these same cars look ever more ridiculous to him. This morning John was listening to another radio report of the flooding and damage along the south coast and caught himself angrily blaming the huge four-wheel drive in front. The only time he feels smug about driving now is when he sees one of the ever-growing number of ex-military personnel carriers belching exhaust fumes around town.

John Carbone used to like going to the gas station. Though the bill for filling the tank stung a bit, it was always a joy to walk back to his big car after paying. Now he hates filling up the car

– he gets embarrassed at the shear volume of fuel it swallows. Watching the dials click round takes so long that smaller cars pull up, fill up, pay and leave while he stands there pumping in more fuel.

The erosion of John Carbone's enjoyment in his SUV reached a new low when Henry, spurred on by a classroom discussion, asked his dad why he drove such a big car to work when there was only him in it? John had no real answer. Some months ago he considered joining a local car pool and at least filling some of the acres of spare seat space. Several people in the street have the same route to work, but sharing really didn't appeal much. They might not like Eric Clapton, or they might be late, not pay their fair-share of the fuel bill, or just be irritating. In the end he had abandoned the idea. Now, with his conscience gnawing at him more every day, the time for some real action has come. Either he'll get involved with a car pool and divide the emissions of his car between more people, or the gas-guzzler will have to go. The latter appeals more: it entails more car research and shopping.

Persuading Kate is easier than expected. She had always felt the four-wheel drive was too ugly, too costly to run, and a risk to pedestrians, quite aside from its emissions. When John explains that his priority is fuel efficiency she actually gets quite interested. Together they go online to look for information. They are shocked to discover just how bad their SUV is for the climate. Compared to a standard five-door hatchback it emits twice the amount of greenhouse gas per kilometre. The hatchbacks they look at seem to have all the same comfort features as the SUV and, by getting a dual-fuel version, John Carbone can cut his fuel bill by three-quarters. Bye-bye SUV.

The Monday morning drive to work following delivery of the new car is one of the most enjoyable ranting sessions John Carbone has ever had. All around him are gas-wasting monsters driven by ignorant buffoons, each sitting by themselves in their climate-controlled boxes. He even signs up for the car-share scheme that coffee-time at work, the glowing feeling of superiority making him grin all day. The fuel station is a revelation. Rolling up to the liquefied gas pump, John notes the enquiring looks from the others on the forecourt and saunters up to pay his small fuel bill.

Each year, this move to a smaller-engined car will slash 6 tonnes from John's total greenhouse gas emissions, equivalent to a more than 50% cut in his total transport-related climate burden.

Despite this great potential for change without any big lifestyle compromises, US transport emissions are expected to increase 50% by the year 2020, to represent a third of total emissions in the USA (they already represent a quarter). If you do nothing else to fight climate change, if it all seems too much effort or just too inconvenient, then at least get a

Figure 5 Jacob Marley and Ebenezer Scrooge discuss engine sizes (many thanks to Paul (Jake) Jacobson).

smaller-engined car. In the next life Dickens' Jacob Marley had money boxes hanging around his neck: how would you like a 4 litre engine?

There are lots of other ways to cut car-related greenhouse emissions. For a start there are a growing number of petrol alternatives – there's big money to be made, oil won't last forever and getting at what is left is proving fraught, to say the least. At my local petrol station there are three standard pumps at each bay, a green one (lead-free petrol), a red one (bad old leaded petrol), and a black one (smelly, sooty diesel). You'd assume that diesel would be the last choice for a climate-aware driver: a litre of diesel produces about half a kilo more greenhouse gas than a litre of petrol (about 2.7 kg compared to 2.3 kg). But modern diesel engines are roughly a third more efficient than petrol engines of equal size and diesel contains about 12% more energy per litre than petrol, so you get almost half as much distance again. Overall then, modern diesel vehicles tend to have between 5 and 10% lower greenhouse gas emissions than their petrol-powered cousins. New particulate filters, low-sulphur fuel and catalytic convertors are really starting to make diesel a viable option for the climate-aware driver.

Back at the petrol station, aside from the air and water pumps, the news stand, and the instant barbecues, you'll probably spot the odd-looking dual-fuel pump. Dual-fuel gas usually comes as CNG (Compressed Natural Gas) or LPG (Liquified Petroleum Gas), with the petrol relegated to a supporting role. It costs much less than regular unleaded petrol – commonly half the price – and produces up to 20% less greenhouse gas per kilometre. Most cars can be converted easily to run on dual fuel. The extra tank takes up about the space of a spare tyre and the cost is recouped in cheaper fuel bills within a few years.

I drive a Smart car – much less engine to lug around in the afterlife. Initially I hankered after one of the new hybrid cars that combine standard petrol power with backup from an electric motor, charged by the brakes. The petrol engine powers the car on the open road, when its efficiency is greatest; the electric motor helps when you're accelerating and takes over completely when you're braking or standing at the traffic lights waiting for green. These cars emit 30–40% less greenhouse gas than similar-sized petrol models. But they remain pricey and on my jobbing scientist's salary there is no way I can afford one. All these can reduce our greenhouse emissions, but all rely to some extent on burning fossil fuels and so will never be entirely climate-neutral.

Electric vehicles have been around for a long time and appear to provide potentially bigger cuts in emissions than hybrid vehicles and the rest. Yet the electricity used to power them is where they can come unstuck climate-wise. If it can be provided by renewable sources, such as wind or hydroelectric, then emissions are indeed vastly smaller than fossil-fuel based alternatives. But most people's electricity is generated using fossil fuels and so by plugging a car into the socket each night they are simply shifting the emissions away from their exhaust pipe and over to the coal-fired power station.

All-electric vehicles also tend to perform somewhat sluggishly. I have no interest in being able to accelerate like a jet fighter, or for my car to break the sound barrier, but when I'm running late and my daughter has tired of drawing snot pictures on the window and opted for all-out screaming, a top speed of 40 mph could get frustrating. Range too is currently rather restricted, being only about 40 miles between recharges in most models. With oil reserves plummeting and fuel prices rising, a switch to electric vehicles is increasingly likely. Their performance will improve and, for around town at least, they should become a very useful option. Still, the origin of the charge in their batteries is key to their success.

Biofuel is another growing, oft mistrusted, option that has the potential to provide climate-friendly driving. Occasionally newspapers run stories like 'Rooster Booster: man in New South Wales converts truck to run on methane from chicken manure', 'Chip Charged: Burnley chip shop owner tops up tank with old oil from pans', and 'Green and Drunk: cross-eyed Texan corn farmer swears home-made moonshine fuels truck and improves driving'.

Biofuels are either crops, like corn, made into alcohol (ethanol), or are byproducts of some other process, like manure methane or chip-shop oil. In theory, the carbon dioxide released when such fuels burn is offset by that taken up by growing more fuel crops, making the fuel effectively carbon-neutral.

In reality, the benefits are not quite so straightforward. Growing biofuels can itself lead to substantial emissions – nitrogen fertilisers for instance can produce large amounts of the powerful greenhouse gas nitrous oxide when soil bacteria use the nitrogen instead of the plants it's intended for.

After accounting for all the production-related emissions of growing, transporting and processing, most biofuel crops end up giving a cut of 20–30% compared to petrol – not climate-neutral, but still very worthwhile. The savings really shoot up when biofuel is made from the byproducts of other processes. Ethanol made by fermenting waste pulp – cellulosic biomass – from paper mills have zero greenhouse emissions.

Forecourt pumps offering biofuel, usually a mix of ethanol (known as E85) or methanol (M85) and petrol, are on the rise. In the USA there are already over two million flexible fuel cars on the roads, able to run on these fuel mixes. In Brazil, a country rich in cane sugar-derived ethanol, sales of alcohol-fuelled cars now top a quarter of a million a year. With more research, in laboratories and people's backyards, biofuels will get better. Fried fish fat may not be the final answer, but alcohol, so long a lethal mix with driving, could see climate-friendly car use take off on a large scale.

Hydrogen has been touted as the solution to car-driven climate change. As hydrogen is used up it produces nothing more harmful than water. Unfortunately, producing hydrogen in the first place tends to burn more fossil fuel than that saved by using it instead of petrol – not exactly the climate change panacea.

In the USA, over a billion dollars is being spent on improving this technology, where it is seen as a way of insuring motoring lifestyles against the uncertainty of oil supplies. The focus of this research is improving hydrogen fuel cells – batteries that sit in your car and convert hydrogen to electricity. The vision is one of US roads filled by these so-called Freedom Cars. The reality is that even with another 15 years of aggressive research and development their climate benefits will be no better than the hybrid cars already available.

Finally, solar power may still have its day, either as a booster to more conventional engines or, where the climate permits, a wholly emissions-free way to power a car. The potential of solar power is clear when you look at something like the World Solar Challenge, where teams from across the globe compete to see whose solar vehicle can cross Australia from Darwin to Adelaide in the shortest time (about four days is the aim). Doing this baking 3,000 km journey in four days powered only by the Sun is certainly impressive, but then Australia is not short on sunshine and the race rather grinds to a halt at night.

Solar energy's role may actually be more indirect. It has great potential to provide all that raw hydrogen needed by the fleets of whirring Freedom Cars and to make this and similar fuel cell technologies a realistic way to cut greenhouse emissions.

It's quite exciting to imagine what might actually make it onto our service station forecourts in the next few years. Will more biofuel ethanol pumps pop up beside the standard red, green and black fossil-fuel ones? Maybe manure-methane pumps will become standard, with a bright yellow chicken logo and a free egg with every cubic metre sold.

Meanwhile, by using dual fuel or, where possible, switching to biofuels, we can make anything from 20% to 100% cuts in our transport greenhouse emissions and encourage the spread and development of these new fuels.

Changing one's driving habits is probably the easiest of climate-friendly actions in theory. By driving habits I don't mean whether you like to wear shades and a baseball cap or have chemical toilet-scented air fresheners dangling from your rear view mirror, but when you use your car and how you then drive it. My bookcase holds several books on individual action, each containing top tips by which to cut your emissions through changing your driving habits. On paper these make absolute sense and every time I read them I nod and think "Great, that's easy – drive smoothly, save fuel and save the planet – no problem". But, such good intentions tend to evaporate when the first car accelerates to within a gnat's tongue of your rear bumper and flashes its lights. It takes real strength of character to cast off driving habits established over decades, but the benefits to the climate, your fuel bills and the safety of yourself and others on the road are undeniable.

The number one tip is to avoid short and unnecessary car journeys in the first place. As with the Carbones' dog, Molly, lots of short car trips can add up to a lot of emissions in a year.

In the UK cars are used for over 60% of journeys of between just 1.5 and 3 kilometres; and for nearly 20% of trips under 1.5 kilometres – a distance most of us could walk in less than 15 minutes. Popping down to the local shops twice a week for that oh-so-crucial extra packet of snacks or vital goat's milk for the dinner recipe can release a third of a tonne of greenhouse gas over the year. While it may be hard for the shopping bag-laden mother to see the attraction of public transport, we are all in real trouble if we can't avoid using our cars by using normal butter rather than popping out to find the last pat of organic hand-made slightly-salted butter in the county.

Out on the open road, speed really is king if you want to cut your greenhouse emissions – assuming for once you aren't too late and can drive pretty much at the speed you want. Instead of racing up to your normal 10 kph over the speed limit (if you usually drive faster then the benefits, both to climate and your life-span, of cutting your speed are even greater), get the car into top gear, put on some classical music and try gliding along at 10 kph under the limit. It feels odd at first, and all the cars whizzing by in the outside lane make you feel like you're going much more slowly. Try this as often as possible over the next week – you will annoy some of the still-speeding drivers, but you can always employ the universally recognised signal of conservative driving by wearing a hat for these trips. For some-one commuting 20 km into work and back 5 times a week the more efficient driving speed will make the tank last about a week longer. In a year of commuting this saving is equivalent to two full tanks of petrol and cuts over a quarter of a tonne off greenhouse gas emissions.

There are a whole range of other fairly common-sense things that help cut emissions. Car air-conditioning can soak up 10% of fuel, so use the vents or windows instead. Keeping tyres well inflated, the car properly serviced and its air filters clean, and not over-filling the tank are all rather dull but worth doing, together saving up to 50% on fuel use. Avoiding sharp acceleration or braking can also save a lot of fuel and so greenhouse gas emissions. Those bunny-hopping tail-gaters that infest our towns and cities are burning their way through an extra 40% of fuel in their quest to shave two milliseconds off their journey. Next time that white van screeches up behind you, adjust your hat, turn up the Bach and pity their ignorance.

The average number of people per car is currently about 1.2 in the USA, adding up to about 10 trillion 'empty seat miles' each year. Increasing car-sharing even a bit could take many thousands of cars off the road. Joining the car pool now cuts John Carbone's car use by two days a week and takes around half a tonne off his annual greenhouse gas emissions. That's another 15% off John's original car-based emissions.

Sadly, several valiant efforts by politicians to encourage car sharing have had rather disappointing results. In Southern California the introduction of high-occupancy vehicle lanes (HOVs), where cars using these fast lanes had to have at least one passenger, pushed up average occupancy from 1.22 to just 1.25. Government schemes, both the carrot kind, like giving cars with passengers their own lanes, or the stick kind, like charging those without passengers, tend to fail because of cheating by drivers. In the case of the HOV lanes, one woman conned police for nearly a year by using a dummy passenger. An undertaker prosecuted for his use of the HOV defended his actions saying that the body he was carrying counted as a passenger, while a pregnant mother claimed a similar defence, her unborn child this time being the 'passenger'. The undertaker lost, but the mother-to-be won her appeal.

To achieve real change each of us needs to see the true incentive behind changing our driving habits – that of climate change mitigation – before it is too late and floods carry off our cars, passenger dummies and all. See Table 1 for a quick comparison of the ways to cut emissions from car use.

Table 1 At-a-glance ways to cut greenhouse gas emissions from cars (savings in comparison to a 4 litre petrol-driven car).

	Smaller engine	Dual-fuel car	Hybrid car	Biofuel car	Driving habits
Greenhouse gas savings	Up to 75%	20–30%	20–40%	Up to 100%	Up to 50%

What about not travelling in the first place? It's something most of us have hankered after at one time or another. The idea of working form home seems an absolute winner when you're running late for work and a day in the stuffy office beckons, interspersed with long meetings about how to increase productivity, or when the traffic jam apparently starts at your front gate and that extra glass of Rioja from the night before is giving you supersensitive hearing and a tongue like a pub carpet. There are very few jobs these days which don't comprise a fair amount of paperwork and email. Some could probably spend only one day a fortnight going through administrative work at home, while others could swap full-time desk-jockey jobs for a decent computer and home Internet connection – as long as they have the willpower to resist reruns of *Cagney and Lacey*.

The climate benefits of home working are potentially great. The trend in recent decades has been for more and more of those living in rural areas to commute into the town or city for work. It's now common for office workers to make a 300 km round trip each day, so allowing them to live in a large house with a garden they never see in daylight. In the USA the average commute has gone up by a third since 1987.

A day without the office run can clearly make substantial reductions in our greenhouse gas emissions. Yet there are real dangers of pollution swapping here. Let's say you convince your boss that working 600 km away in an Internet-ready log cabin will increase your productivity and be a veritable gold-mine for blue-skies thinking. You make the big move to the sticks and, apart from the problem with your electricity supply going off whenever the wind gets above a strong breeze, all works well. No more of the daily 160 km drive/crawl to work and back. So far, so good. However, you still need to go into the office occasionally for special meetings – mainly because the best thing about working from a desk in a log cabin, with a view of a limpid lake and forested mountain, is being able to tell your colleagues how great it is.

So, you keep on a flat in the city for those trips back to the office, which average about one a week. The flight down and back clocks up just over 1300 km, almost 500 km more than you used to cover each week driving to work and back, and using a more climate-warming mode of transport to boot. The city crash-pad itself means an extra set of appliances and fur-nishings, all of which use up yet more energy in their manufac-ture. Through your 'climate-friendly' home-working, therefore, your related greenhouse gas emissions leap from 112 kg per week to about 180 kg.

Other pitfalls lie in the extra energy we use at home. Instead of the daytime being a period of low home energy use, we are now in the home, requiring air conditioning, illicitly watching TV and, usually at about 2 p.m. craving some human interac-tion and driving to the local garage for a newspaper and half a gallon of petrol. With two offices, one at work and one at home, we have all that energy used to make two desks and two PCs, two printers and two pen holders. At the same time, the energy for heating, cooling and lighting your office at work is still burning away, whether you are there to appreciate it or not. Home-working can potentially end up producing more greenhouse gas than sticking to the grind of the office run. The solution then, is to use the technology available in a way that makes a tangible difference to your climate impact: virtual conferencing for example, instead of swapping five short commutes in the car for one big one in a jet plane. Which neatly brings us to flying.

Flying these days plays a big part in nearly everyone's greenhouse gas budget. Air travel is already a huge climate problem and it's growing fast as flights get cheaper and more frequent and go to more destinations. A return flight from London to Sydney spews as much greenhouse gas into the atmosphere for every passenger as that given out by a good-sized car over an entire year (over 4 tonnes). The United Nations predicts that, by 2050, annual aircraft emissions will have reached well over a billion tonnes of greenhouse gas.

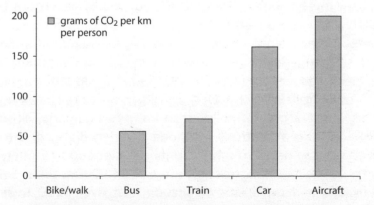

Figure 6 Average carbon dioxide emissions for various modes of transport.

The climate-warming effect of air travel is magnified even further by the fact that most of the emissions occur high in the sky, at the very place where they can do most harm. As well as pouring out tonnes of carbon dioxide planes produce large amounts of so-called NOx gases (oxides of nitrogen). Up there in the troposphere, these NOx gases lead to the formation of yet another powerful greenhouse gas, ozone, effectively doubling the climate warming effect of our flights. Had we deliberately set out to cook our planet we couldn't have done much better.

Given its big role in global emissions you'd expect air travel to be feeling the pinch of efforts to tackle climate change. Not a bit of it. Because planes trickily fly backwards and forwards across international boundaries their emissions aren't included in national budgets under the Kyoto protocol. At the same time, aviation fuel goes untaxed and the industry benefits from large government subsidies. So we get a burgeoning budget airline industry, with seat prices less than the taxi fare to the airport and the true environmental costs of jet travel hidden away under the dirty aisle carpets.

Our love of air travel is going to turn into a rocky affair as global warming intensifies. Baking summer temperatures will attack runways, just as they attack our roads; and more frequent and intense storms will ground increasing numbers of planes. When it's not too windy and the runway is OK, planes may still struggle to get airborne with soaring summertime temperatures meaning less lift. Weather is already the cause of 70% of flight delays in the USA, causing a quarter of all air accidents and, through flooding, storm and hail damage, closing whole airports at a stroke. Just one diverted flight, due to a storm say, can delay 50 other flights and clock up $150,000 in costs. Cancelling a flight costs around $40,000 a time. Such diversions and cancellations already cost over $250 million dollars each year.

Where we actually want to fly to will also change. Many traditional summer sun destinations are facing failing water supplies, scorching summer temperatures and a host of disease outbreaks. Winter resorts in some countries are suffering from a decided lack of winter. In Scotland the ski industry is tottering on its last tartan pole, with the number of ski days having dropped by a quarter in the past 20 years.

With huge, climate change-fuelled increases in emigration, illegal immigration could soar. The political instability expected to arise from global warming may rule out some currently very popular destinations altogether, while increasing safety fears will lead to longer security checks.

All of which makes jetting off for that traditional getaway look less appealing. The benefits can be big to the climate of foregoing the dubious pleasures of sticky-carpeted departure lounges, intimate strip searches, and sitting on an overcrowded plane next to a time-share salesman with questionable personal hygiene. For an average long-haul flight, every kilometre you fly clocks up about 150 g of greenhouse gas. A flight from New York to London emits three-quarters of a tonne of greenhouse gas into the atmosphere for every person on board. Short-haul flights belch even more per passenger per kilometre because of the extra fuel used to haul those tonnes of plane, sweaty salesmen and limp egg sandwich into the air.

Rail or coach travel can radically cut emissions on journeys under 800 km. An average short-haul return flight from Amsterdam to Munich emits over 100 kg of greenhouse gas for each passenger; the same journey by train or coach results in emissions of just 30 kg.

As holiday travel makes up over 60% of air travel, rediscovering the delights of vacations in your home country can pay big dividends. Let's return to the Carbones.

It's going to be an odd summer. Seemingly forever now the family has flown *en masse* to Mexico for two weeks of blissful sun and relaxation. With Kate Carbone working for a travel agent they have always got fantastic deals on their flights and stayed in a wonderfully rustic farmhouse about a kilometre from the sea. Now though, for the first time in six years, the Carbones are looking elsewhere. Over the past couple of years the temperature has been unbearable. Even at night it was so warm that just sleeping was a challenge. Last year they had to go without showers for most of the holiday as water supplies hit an all-time low – the final straw for John and Kate.

Kate noticed from the holidays she had been selling over the past few years that Mexico was becoming much less popular during the summer. There were complaints and the tempera- ture guidelines in the tour brochures had been amended upwards three times in as many years to cover the record- breaking summer heat. At first the company policy had been to say that the hot weather had been 'exceptional', but it was becoming obvious that ragingly hot summers were now the norm. So, no Mexican holiday this year. Instead, the family are consulted on an alternative destination. This is a great excuse for lots of arguing, re-reading of travel brochures and the making of unreasonable demands. Once the idea of trips to Antarctica (Henry), Afghanistan (George) or China (John) are all dismissed on the basis of price/danger/distance the family agree that what they all would really like to do is to go somewhere cool, with good food, swimming, and no flying. After a few phone calls to the office Kate confirms that their new holiday is booked. They're going down south to Tickfaw, Louisiana, to a lakeside cabin just north of New Orleans, for a winning mix of swimming, walking and, if John has anything to do with it, jazz.

When holiday time finally arrives, the Carbones are feeling ready for a break. The weather has been dry and hot and everywhere seems to be covered in a fine layer of dust. There is concern on the news that it could turn into a real 'dust-bowl' summer and the Carbones have spent many days dreaming of slipping into the cool waters of Lake Pontchartrain. The packing is the usual frenzied affair, with the added complication of a new holiday destination and so a barrage of questions about what might and might not be needed. After Kate has drawn the line at taking a TV 'just in case' and made clear to John that, no, there isn't room for him to take his entire Miles Davis collection to get himself in the mood for New Orleans, they are ready to go. After three false starts (for, respectively, a forgotten pair of sunglasses, another toilet stop for Henry, and retrieval of an errant Labrador), the Carbones are off.

This switch of holiday, from Mexico to Louisiana, swaps the two tonnes of greenhouse gas produced from the family's return flights to Cancun for less than a third of a tonne – Kate Carbone's barn-like people carrier is for once full of people.

In the USA there are 405 million long-distance (more than 80 km) business trips made each year, accounting for 16% of all long-distance travel in America. For business meetings less than 400 km away most people opt to drive, but as soon as the travelling distance gets up past the 800 km mark air travel is the usual choice.

For most international journeys air travel remains the only practical option, so what then? Several schemes have sprung up to massage our emissions guilt. These generally involve paying for trees to be planted or making contributions towards new renewable energy schemes to offset the emissions from your flight. During their 2003 tour, the Rolling Stones took their responsibility for greenhouse gas emissions to a new level. They paid for enough trees to be planted to offset the emissions of each one of the estimated 160,000 fans attending their UK gigs. At 13 kg of greenhouse gas per fan this meant a lot of trees – 2,800 to be exact. Great, but imagine the number of trees it would take to offset every tour these long-lived rock legends have ever made. Between 1997 and 1999 alone, over 5 million people went to see the Stones at 147 different gigs across Europe, That's another 100,000 trees.

It's unrealistic to think that all our air travel emissions can simply be offset by planting trees. The primary worth of such schemes is to get people thinking about their climate impact, rather than to solve it. Some schemes – 'ClimateCare' for example – now allow you to offset your emissions not with trees but through funding renewable energy schemes in the developing world. The climate benefits of doing this may be more transparent, but there's no easy way around this. To minimise your contribution to global warming fly less, or not at all.

As a scientist, one of the 'perks' of the job is going to international conferences. These are usually dominated by people with a penchant for wearing socks with sandals and for jokes about the time in '78 when Barnaby Lorimer-Smyth put up a slide of an anterior section of *Biterson vulgaris* upside down and didn't notice. They are part of every academic's working year. The premise is that they allow us to present our latest research and spark off new ideas in an atmosphere of convivial academic review. The reality is that the research presented is usually several years old and has already been published, and that most people go to network (young, dynamic and, most of all, desperate-for-funding researchers), or to meet up with old friends (socks and sandals). Occasionally, these conferences do work well and can get some useful discussions going. However, there is a tendency for them to be seen as jollies – free holidays paid for by the taxpayer to make up for a researcher's poor pay and working conditions. I'm sure it applies to business too. Barnaby Lorimer-Smyth may instead be Harris Baines, socks and sandals may be replaced by a novelty Simpsons tie, and the upside-down slide joke may instead be about Baines eating the boss's bread roll, but it all boils down to an awful lot of air travel.

Cutting greenhouse emissions due to conferences and meetings can be as simple as having them in the right place. For a
meeting of 10,000 scientists in San Francisco (many of them climate change researchers) a couple of years ago it was estimated
that associated emissions for the meeting topped 12,000
tonnes, with the average delegate travelling about 8,000 kilometres. Had the conference instead been held in the more central location of Denver, Colorado, the emissions would have
been cut by 900 tonnes.

As well as getting the place right, you can slash emissions by
attending meetings virtually. By avoiding just two medium-
distance flights of, say, 1,600 km each way, you'll cut nearly a
tonne of your annual greenhouse gas budget.

Several international conferences have already been run virtually. A recent genetics conference in the USA cut greenhouse gas
emissions by about 900 tonnes through delegates swapping red-
eye flights, soulless hotel rooms and complimentary soaps for
their own office and the ability to go home for dinner each evening. For small island states there is now the 'Virtual World Forum
on Small Island Developing States', an obvious use for this technology given all the water separating the interested parties. (The
theme for Session 1 of the next meeting? Climate Change.)

The standard format of these virtual meetings is for the
delegates from each place to get together in their own conferencing room and then, using a large screen, cameras,
microphones and the power of the Internet, talk to groups in
other cities, countries or continents.

On the large screen is the main speaker doing his or her
thing, while windowed at the side are the other groups waiting
their turn to speak. This high-tech conferencing allows you to
download whole presentations – all the facts and figures – to
peruse at your leisure. It also means that if anyone asks a particularly difficult question you can frown, point at your ears, and
mutter something about Microsoft.

Real conferences will always have a place, but, given the many millions of air kilometres clocked up by delegates each year, even limited use of virtual conferencing has the potential to reduce greenhouse gas emissions by many thousands of tonnes.

Transport, then, plays a leading role in our individual contribution to global warming. But before you doze off to dreams of being run over by an SUV driven by a sandal-wearing chicken it's time to turn off the car-choked road and into the drive. It's time we took a look at the impact of our homes and, most worryingly, how climate change is planning to visit.

3

home start

If you live close to a river, near the coast, or in a low-lying area, you may want to look away now. Each year, floods affect more than half a billion people and claim around 25,000 lives. Currently the United Nations designates about 1 billion people 'at risk' from severe flooding worldwide. Predictions are for this number to double during the 21st century as global warming intensifies storms and torrential rains. What's in store are not Hollywood-friendly new types of weather, but more of the existing devastating ones.

To get a good idea of how big the threats that our homes face are we only need look at house insurance. Insurers are essentially gamblers – well dressed, polite gamblers. Their livelihoods depend on correctly estimating the risk, say, of your house falling into a mineshaft or your chinchilla needing dental surgery. To refine their risk assessments they use the best information they can find, from detailed mining maps and surveys to in-depth studies of chinchilla eating habits and longevity.

So how have these cautious folk responded to climate change predictions? By offering soaring house insurance premiums to many and, to those in flood-risk areas, no insurance at all. Companies around the world are hiring climate experts, frantically drawing up new 'At Risk' tables, and upping their premiums. The most pessimistic are moving their head offices away from swanky docklands areas to locations well above sea level.

In the USA, 10 million homes are already at risk of flooding and the flood damage bill tops $7 billion a year. With sea levels set to rise somewhere between half a metre and a metre by the end of the century, 36,000 square kilometres of the USA – almost twice the area of New Jersey – may be lost. According to the US Environmental Protection Agency such flooding would cost between $270 and 450 billion. Insurance claims are likely to double or triple, and thousands of families could be displaced.

In Eastern Canada sea level rise could be even more devastating because the land itself has been steadily subsiding since the last ice age (about 2 cm a decade in some areas). In Australia the potential for displacement of communities is huge: 80% of the population live within just 50 km of the coast. In the UK, more than 3.5 million homes could flood this century – three times the number currently under threat. Globally, coastal flooding due to sea level rise will be threatening an extra 23 million people by 2050.

When the flood waters finally subside and that special post-flood smell – from a dilute mix of petrol and sewage – begins to disperse, at least it will be good drying weather. The predictions of temperature rise this century are big. Here in Scotland I've got my grapevine planted in anticipation of a Chateau Reay Rioja. Imagine a hot summer now, a real belter, say the hottest in ten years. In the Europe of 2080 nearly every summer will be just like that. Add to this a 30% drop in rainfall and the fact that summers are getting warmer and drier in the north, and it's bye-bye to the great wines of Bordeaux and hello to the cheeky reds of Brittany, Bognor or even Bute.

Summertime in New York by 2050 will be more like that currently experienced in Atlanta. In Atlanta, summers will be like those today in Houston. And in already-hot Houston? Think Panama (temperatures in the high 30s and close to 100% humidity). Heat stress already kills around 400 people in the USA each year. Europe had a taste of what is in store when, in 2003, over 20,000 people died as a result of an intense heatwave.

A table in the weighty volume *Climate Change 2001: Impacts, Adaptation and Vulnerability* – produced by the Intergovernmental Panel on Climate Change – sets out what climate extremes are coming our way. Below the heatwave section ('increased incidence of death and serious illness in older age groups and urban poor'), are similarly stark predictions for 'intense precipitation events' (heavy rainstorms), 'increases in peak wind and storm intensities', and 'intensified droughts and floods'. These worsening extremes will mean more damage to buildings, infectious disease epidemics, property losses and greater risk to human life.

Climate change will also have some less obvious impacts on homes. Already, extreme summertime temperatures and droughts have led to a big increase in subsidence as soil shrinks and trees suck the last drops of water from the soil. In Alaska thawing permafrost has seen houses literally disappearing into the ground.

All in all, the outlook for our homes is pretty worrying. If you live 2 m or less above sea level, as millions around the world do, then buying that new carpet for the living room might not be such a good idea. If you're on a hill you might want to secure those loose tiles before the next storm, and if you're nestled in a city high-rise make sure you've got a good stock of bottled water and that the air-conditioning is regularly serviced. And order a Panama hat.

Alternatively we could work to keep the greenhouse gas concentrations in our atmosphere below those we're currently speeding towards. In 2004 an otherwise grim report on the threat of flooding in the UK this century pointed out that by reining back global greenhouse emissions by 25% the costs of damage due to flooding could be cut by a quarter (from £21 billion to £15 billion a year in the 2080s).

Homes are massive energy users and so tend to be massive emitters of greenhouse gas. Happily, home is also where some of the most straightforward changes can be made. Take something as simple as swapping an old-fashioned tungsten light

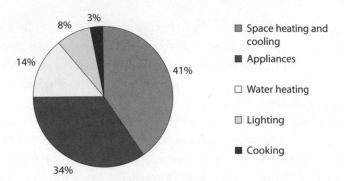

Figure 7 Greenhouse emissions of an average home in the USA (% of 11 tonnes of greenhouse gas per year).

bulb for a low-energy one. Easy to buy, no extra effort to put in, and a money saver to boot. Just one bulb can stop up to 100 kg of greenhouse gas getting into the atmosphere each year. Indeed, almost all the climate-friendly strategies for the home – those listed on government websites, in books, even on the back of your electricity bill – reduce energy wastage and so save money as well as the environment. Making our homes more climate-friendly really can hit that 25% flood-stopping target, and go beyond it, even to the scientists' recommended 60%.

It's the control of our home climate that most threatens the global one. Like some loyal dog with a pair of chewed socks, the heating or air-conditioning system welcomes us with exactly the temperature we asked of it. We've come a long way from the spitting fires in damp caves that first constituted home heating. Gone are the winter nights which would freeze slippers to the carpet, or the summer days when even house-flies were too hot to bother buzzing. Now we can slop around our houses in Bermuda shorts and a Che Guevara t-shirt as the snow piles up against the windows. After some vigorous air guitar to Dire Straits, we can complain that it is too hot, pad barefoot across the shagpile and open the window.

Regarding transport I said that some environmentalists' criticism of car driving was a bit much, given the alternatives available. But with home heating and cooling they can literally pull on their woolly jumpers and get on their high, organically fed horses. I can hear my mother's words ringing in my head as I write: "If you're cold, put a jumper on, David. Nana knitted you that nice one for Christmas and you've not worn it once". Simply putting on more clothes when it's cold, or taking more off when it's hot, can slash household energy use, and so greenhouse emissions. Heating the average single family home in the USA emits about 4 tonnes of greenhouse gas each year; cooling it emits a further tonne. Here in Scotland, heating-related emissions also tend to be big, although air-conditioning isn't widely used – yet. For an average developed-world household, knocking the heating thermostat down just 1 °C, or the air-conditioning up by 1 °C, will cut a third of a tonne off annual emissions.

It takes a real shift of mindset to go from blithely cranking up the air-conditioning in our homes as the sun beats down outside, to turning it down in order to take some of the heat out of the summers to come. That whirring air-conditioning system is like a climate change loan shark: it gives you a cool house today but means a hotter world tomorrow, with interest.

Demand for housing is predicted to increase across the developed world, as more and more people aspire to home ownership, single living and bigger houses. Across Europe the demand for new houses will stay at well over 2 million a year for the next 20 years. In Japan, the number of single-resident households is expected to overtake the number of homes with two or more occupants by 2007, rising by a third in the next 25 years. Already, in Vienna almost half of households are one person only. Over the last 30 years the average number of people in Canadian and US households has dropped from 3.2

to 2.6 – like every two houses in the 70s sprouting into three houses today. At the same time the average house size in the USA has increased from 1,500 square feet (140 m^2) to more than 2,200 square feet (205 m^2).

Building design has become more energy-aware, but the trend is still one of increasing energy use in, and so greenhouse emissions from, housing. Over the next 20 years the combined energy burn in US homes is set to increase by another 20%. A person living alone gets through twice the amount of energy as someone in a three-person household, and nearly five times more than someone shacked up with five others. With more than one person in a house you share heating, lighting, and even that kettle of water for the coffee. Live alone and, though you'll always have the final say on what to watch on TV, you'll get through a lot more energy as an individual. The trend towards bigger houses has the same effect – bigger houses mean more materials, more heating and cooling and so more emissions.

In front of me is a pile of glossy government-sponsored brochures from the USA, the UK and Australia promoting energy-efficient strategies in the home. They each contain pictures of people drawing curtains, putting in loft insulation and closing double-glazed windows, all with rather off-putting self-satisfied grins playing on their faces. Grins aside, the energy savings, and so benefits to the climate, are impossible to ignore. Without exception, the brochures extol the virtues of improved insulation. This can take the form of thicker padding in the roof space, installation of double glazing, or simple draught-proofing of existing doors and windows. Getting cavity wall insulation, improving pipe and boiler cladding, and adding internal doors to provide enclosed hallways, can also prevent heat loss and energy wastage. The numbers are impressive: by better insulating your home you can almost halve the amount of energy used to heat or cool it – a cut of more than 2 tonnes of greenhouse gas emission a year for an average house.

As with transport, governments have recognised household energy use as a key battleground in the fight against global warming. As climate change intensifies we can all expect to see a lot more of those self-satisfied grins in government booklets and on our TV screens. The reach of governments in cutting household energy use can go way beyond posting booklets through our letter boxes and broadcasting 'Is your home behaving badly?' adverts. In some countries, Australia for instance, you may be able to claim back part or all of the expense of better insulating your home. Government-funded housing developments, with energy-efficient building designs, are also becoming more common. The Guguletu Eco-Homes project in South Africa is a good example of this, involving the construction of 6,000 energy-efficient homes over the next 50 years. By reducing energy use in space heating, primarily through solar heating, this project is set to cut emissions by a massive 40–50,000 tonnes of greenhouse gas over its lifetime.

Few, if any of us, bought our current home with climate change in mind. If we had, then all those flood-plain developments would be in serious financial trouble. When we do next choose a house, there are some key climate-related questions worth asking:

- Is the new house closer to work, the station and/or the cycle path?
- Is the house such a rambling, draughty, energy-guzzler that just keeping the thing warm in winter will eat up more power than your current house uses in a year?
- Is there a garden, no matter how small, to grow you own food in?
- Can you get renewable energy in the area?

And so back to the Carbones to see how housing decisions can radically affect our contribution to global warming. There's big news for the whole family and some major decisions to be made which could drastically affect all their climate impacts.

The Carbone family is going to get bigger. John and Kate Carbone have kept the news to themselves for the first few months just to play it safe, but now, 6 months in, Kate is showing clear signs of a bump and they've had to break the happy news to Henry and George. After initial apathy from both, George posed the tricky question of where this new addition was going to sleep. The Carbone house, seemingly so big when they first bought it, has since filled up with the detritus of modern life, not to mention a Labrador and two sons. Having assured both Henry and George that "No, they wouldn't be made to share bedrooms with the new baby", John and Kate have to admit that they will need more room. They aren't quite decided on whether to build an extension or start looking for another house. This decision is one of the biggest they will ever make climate-wise. The extension has a great advantage in that the Carbones could ensure that building design and materials were as energy-efficient as possible, while a new house has the potential to be much more, or less, energy-efficient throughout, depending on its design.

They're in a good position for a move as the housing market is fast-moving and there are several new building developments going up on the outskirts of town. With Grandma Carbone having made a present to them of several thousand dollars, they could stretch to quite a big place, or to a substantial extension. The building option, though, is limited to going upwards into the current roof-space, as anything going outwards would shrink the garden, something that the whole family, Kate and Molly in particular, would find unacceptable. After a long chat the Carbones decide that they'll have a look at a few places while the plans for the extension are drawn up and then decide for certain.

That same weekend the whole Carbone family set out to view the show houses on the new estate. It's certainly a very grand development. At regular intervals on the approach road there are huge billboards extolling the virtues of Hawthorn Hamlet: 'An exclusive development of 3-, 4- and 5-bedroom executive properties set in an unspoilt, secure setting'. Long before they get to the estate they can see its row of shining white flagpoles, each carrying the fluttering emblem of the development company. Drawing closer, the huge new entrance road comes into view bordered by a high brick wall and curving round to where two imposing black gates stand open to reveal the estate within. Once parked, the Carbones pick up the various brochures from the estate office and set off in the general direction of the nearest show house – a four-bedroomed example of the 'Manse' category, 'ideal for the growing family' according to the blurb. The house is certainly immaculate, with the dining table set for a sophisticated dinner, towels neatly folded on the bath rail and flowers on the reproduction mantlepiece. An hour later, the Carbones have had enough of looking at show homes to last them a lifetime. The sameness of all the houses, not to mention the enclave-like setting, doesn't appeal. For a comparable price they can get an older house with the same number of bedrooms, but with substantially more garden and at least some feeling of community.

So the search begins for an existing home in their area, with at least four bedrooms and a good-sized garden as must-haves. After a few false starts the Carbones find a property that seems to fit the bill, only a few streets from their own house. The house is quite old, 54 years old to be exact, and is 'in need of modernisation' according to the seller's details. However, it does have four good-sized bedrooms and a really big back garden which, though overgrown, sets Kate Carbone's eyes alight with its potential and gets the wholehearted approval of George, Henry and Molly. It's only when the survey comes back that the true import of 'in need of modernisation' hits home. The cost of this modernisation is likely to be as much as that of the house itself, and would make living there uncomfortable for a long time. With a new baby to think of, not to mention the bank manager, the Carbones finally plump for an extension.

Climate-wise, the flagpole-heavy new development houses are much more energy-efficient than the 'in need of development' old house. A new house will use about a third less energy for heating than an equivalent house built in the 1950s, mainly due to better insulation. But there's a catch to these hermetically sealed executive retreats: embodied energy. This is the energy that has gone into each brick, tile and faux Georgian fireplace. Indeed, every manufactured thing in our homes, everything that has taken energy to produce and transport, carries with it an invisible climate price tag. The PC I'm writing on now has one. It's using energy to power the monitor, to spin the discs and fans, power the processor, and to play *Californication*, but even before it arrived at my house it had used a great deal of energy. Even before I had torn open its box, unravelled its endless cables and sworn roundly at the author of the installation manual, the manufacture of the PC and its various components had used up the equivalent of 240 kg of fossil fuels.

The same goes for the bricks used to build a new house, the slates, the concrete and even the Welcome Home mat – all carry an embodied energy tag. Take just one brick in your wall: to get that brick where it is today required energy to mine the clay, transport it to the brick factory, cut it to shape, fire it in the kilns, ship it to the building site and then lay it in your wall. For a whole house, therefore, the embodied energy can be huge. On average, a brand new house has already used the same amount of energy as will be used by its owners over the next decade of home-living. As such, it has been responsible for around 70 tonnes of greenhouse gas, even before the occupants make their first coffee.

The case, then, for new houses being more climate-friendly than old ones is not so clear cut. Yes, old houses may be draughtier and waste more energy on heating, but as a house gets older its embodied burden on the climate is stretched thinner, and the energy used by those living in the house comes to the fore. Of the three options for the Carbones – old house, new house or extension – the of an extension wins hands down climate-wise.

As the extension is effectively a new build they are able to make it well insulated and energy-efficient, while keeping the amounts of energy-rich materials to a minimum. Aluminium and steel carry big tags due to the large amounts of energy used in their mining and processing. Timber, on the other hand, tends to be a far less energy-intensive building material. By choosing wood construction the embodied emissions of a house can be cut by 80% compared with those of a steel-framed one, and by more than 85% compared to those for a concrete dwelling. Where reclaimed building materials are used the embodied energies can be reduced still further.

The substitution of timber, or straw bales if you're feeling adventurous, for materials like concrete and steel can therefore make great savings in a building's contribution to global warming. Forests already have a big role to play in sucking up and trapping carbon dioxide, and by making the best use of the wood that comes out of them we can deliver a double blow to the enhanced greenhouse effect.

With the luxury of designing a house comes a wide range of climate-aware options. Long before the building starts, there's something called 'passive design' to consider. This is where the orientation of the house, specific roof pitches, window placements and aspects, can all be used to make the most of solar heating or to minimise the need for space cooling. Simple things, like putting the main living areas on the side of the house that gets the most sun, can mean substantial savings in the energy used for heating and lighting. Then there's the opportunity to use high-specification insulation throughout and to fit doors and windows that minimise draughts and heat loss.

There are also some big energy pitfalls to skirt around with any new build. Most of us, given the chance to have a house built for us, will want something big. When we're given the chance of, say, a bespoke kitchen, it suddenly takes on the guise of John Carbone's SUV – a showcase of our success, all gleaming chrome and granite worktops. Plans end up featuring a kitchen-diner with an 'island' the size of a billiard table, enough storage to house the food for a polar expedition, and a table that can comfortably accommodate three Last Suppers. Big kitchens, living rooms and bedrooms mean big houses. This means a lot more concrete, bricks, steel, and embodied energy. Bigger houses also take more lighting, heating and cooling. A climate-aware new build will therefore combine energy-efficient design and low embodied-energy construction to make a house that doesn't share its dimensions with an aircraft hangar.

The house we live in plays a big role in our climate impact throughout our lives. There's a chance to radically change this for better or for worse when the kids have finally left home, you've drunk your last office coffee and collected your carriage clock. It is at this stage that so many of us up sticks to move to the coast/closer to the grandchildren/further away from the kids. Like Grandma Carbone.

She's now moved into the retirement community apartment, which has a suite of rooms, including a small kitchen. As well as being close to the family it also has wonderful lawned gardens all around. Her decision to sell the old house and move into the apartment pays big climate dividends. The apartment makes great energy savings by being part of a larger building, her neighbours act like layers of the very best blue-rinse insulation. Just keeping the old house warm in winter and cool in summer used to get through almost four times the energy it now takes to keep her apartment the right temperature. Overall, her house move has meant a cut in her home energy use and related emissions by a whopping two-thirds.

Appliances in the average home are responsible for over 4 tonnes of greenhouse gas a year and threaten to overtake heating and cooling as the biggest source of home energy use. Simple message then: always buy the model with the highest efficiency rating. This isn't the whole story though. Rushing out this weekend to get a shiny new refrigerator to replace a slightly less energy-efficient one ignores that old enemy embodied energy.

Tonnes of plastic and metal go into making our appliances. Before switching a three-year-old dust-entombed fridge for a brand new, cobweb-free one, it's usually worth giving the thing a dust-off (just keeping the coils and door seals clean can cut 200 kg off its greenhouse emissions each year) and getting a few more years of use out of it. Sure, if we're fitting out a kitchen with the necessary appliances for the first time, then opting for the most energy-efficient version is a must climate-wise. But most of us already have a kitchen complete with glossy-fronted oven, refrigerator and washing machine. In this case, the longer your refrigerator or washing machine keeps going, the more it is effectively running down its embodied energy, eking it over more and more years. As a rule of thumb, if it's less than five years old and still working fine: keep it. With new appliances, as with designing that new house, the trap of 'bigger is better' is waiting for us again.

Kate Carbone is excited. The kitchen table is covered in glossy brochures and magazines filled with pictures of stunning kitchens invitingly lit by spring sunshine streaming through leaded windows. Other, rather less evocative pictures, show blinding white refrigerators and washing machines, along with sheet after sheet of specifications on why this model happens to be the biggest, fastest or whitest. The Carbones are having a kitchen overhaul with the cash saved from extending rather than moving. Their current kitchen hasn't changed much since they first moved in, though over the years various new appliances have come and gone, including their first washing machine, which took exception to a full load of nappies, threw itself across the kitchen floor and caught fire. Now, a combination of the Carbone dishwasher making the dishes marginally dirtier than before they were put in, Christmas coming up, and so many TV makeovers of kitchens, has galvanised them into action. For his part John Carbone is interested in how big an oven can be bought and how many hobs they will have, though in the finer points of the new kitchen design he is less enthusiastic.

The Carbones pay a visit to the kitchen design desk, set amongst a maze of show kitchens at the local mall. Within minutes, George and Henry are foot-stampingly bored, finding it difficult to get excited about brushed steel extractor hoods and Victorian filigree cupboard handles. After ten minutes of moaning they are packed off to the nearby Gamezone with strict instructions not to buy anything and to come back within half an hour. Kids despatched, the Carbones sit down with the fast-talking sales assistant who, after getting them to answer a series of questions, produces a set of plans and 'artists' impressions' of what their glorious new kitchen would look like. The picture even has a couple, just like the Carbones, apparently sipping wine as they admire the beauty of the work surfaces. The Carbone sons, though, seem to be absent.

Having talked at length about 'the magic kitchen triangle' and the best size of sink, they get on to the choice of appliances. The options on offer are dazzling: hobs with eight rings, ovens big enough for half a horse, let alone the Thanksgiving turkey, and computer-linked refrigerators able to keep track of its contents and let you know when a re-stock is required. The cost is equally impressive, with replacement of all the Carbones' appliances adding up to about three times the price of the kitchen makeover. In the end they head home promising to phone the salesperson at the first opportunity once they've had a chance to think about things. After prising George and Henry off the games console in the shop next door, the jaded family head back home loaded with yet more brochures, plans and specifications.

It is only when John and Kate finally slump down at the kitchen table with a steaming cup of strong coffee and stare once again at the pile of brochures that they start to voice their worries about the mass dumping of their not-so-old kitchen appliances. OK, the dishwasher may be misbehaving, but John is pretty sure that with a clean of the rotors it would be as good as new. As for the other appliances, they may not be as big or gleaming as the dozens they have looked at that morning, but they do the job. It may be fun to have a talking refrigerator – well, for the first five minutes anyway – but their current one is only three years old and is in perfect working order. The fleeting image of dumped refrigerators in hedgerows, a sight they are both sick of seeing when walking Molly in the local parkland, adds further weight to their feeling that such revolving door appliance buying is a bad idea. A giant oven and a hob the size of a table tennis table would certainly be good for those family feasts, but they've always managed in the past with their standard-sized oven and its four rings. Kate then confesses that she was only ever really interested in getting the rather beaten-up looking cupboard fronts and worktops replaced and that it seemed almost criminal to throw out what were perfectly good kitchen appliances to make room for another cutting-edge set which would remain at the cutting edge for all of a week.

That afternoon, with the help of a kebab skewer, John sorts out the dishwasher problem and, after a successful first run, the Carbones decide for definite that they'll keep all their current appliances and just revamp the doors and work surfaces. As John goes off to phone the shopping mall and break the bad news to the commission-hungry sales assistant, he can hear Kate explaining to a rather disappointed Henry why they really don't need a refrigerator that tells them when they need to buy more juice.

Given the Carbones' existing appliances aren't all that old and already have quite good energy ratings, the large new ones they were planning to buy actually come out as greater energy users than their older cousins. The big fridge, though at the cutting edge of technology, will produce three and a half tonnes of greenhouse gas over its lifetime, compared to just over two tonnes from the smaller, less talkative, Carbone fridge. This is in addition to the embodied energy penalty carried by the new appliances.

At some time, things do have to be replaced. Just one too many slices of jammy toast will get posted into the video recorder, the washing machine will rebel against having to spin-rinse enough beach towels to make a Bedouin tent, and the fridge will take umbrage at never being cleaned and start running at –20 °C. This is where those energy efficiency ratings come into their own. Big energy-using appliances in our homes include boilers, refrigerators, washing machines and, for more and more of us, air-conditioning. With most of these, simply not having one is not really an option. Instead we are faced with the choice of model, make and size, all of which can have a substantial effect on the contribution our households make to global warming.

Energy advice labels appear on most new appliances. The benefits of going for the more efficient models can be big. The labels vary from country to country with a star rating (the more stars the better) in the USA and Australia, and a letter rating in the EU. For letter ratings, A is the most efficient, dropping to G (for godawful). All allow you to compare one model with another and so make an informed purchase. If you're the proud possessor of an A-rated washing machine you'll be cutting up to a third of a tonne off your annual greenhouse gas emissions compared to someone with a G-rated energy-guzzler. Even bigger savings are to be made with boilers. New generation condensing boilers have great fuel efficiencies (around 90%) and can take more than two tonnes off the greenhouse emissions from home heating. If everyone in the UK with gas central heating installed one it would save enough energy to heat 4 million homes, and cut greenhouse gas emissions by 17 and a half million tonnes.

Aside from the standard household appliances that are big individual energy users, there is a wealth of other less essential electrical appliances and gadgets common to most of our homes. Every one of these gizmos carries a cost to the climate, used or not (embodied energy again).

On the worktops in our kitchens, the shelves in our living rooms, and even the window sills in our bathroom, there's stuff. Lots and lots of stuff. As good members of a consumerist society we have bought into the idea that we must have more stuff, until we either need to put up more shelving, have a gigantic garage sale, or move to a bigger house. Countless tonnes of infrared binoculars, pasta-making machines ("Make your own fresh pasta in only five and a half hours") and nose-hair trimming sets ("Articulated arm reaches even the most hard to get to areas") now lie forgotten and unused in the houses and self-storage warehouses of the developed world.

Christmas is our bonanza time for sending and receiving these gadgets and novelty gifts. It is, after all, the time for giving, and what do you get those friends and relatives who already seem to have everything? With the shopping days until Christmas rapidly falling to single figures it's either an online shop to see if the executive toys website can deliver in time, or its back to the simmering mix of sweat, tears and *Jingle Bells* that is the shopping mall.

In the end, most of us are just glad that we managed to get the last Portable Walrus Cleaning Kit and it's warm smiles all round when the big day comes and we exchange our hard-won present for our very own Horse Grooming set, with fetlock volumiser. Clearly, demanding that all your friends and family refrain from buying you anything other than clearly specified items is something which gets increasingly difficult from the age of about 8 onwards. Each of us can, though, reduce the unwanted 'stuff' problem by buying presents for others with a much smaller climate price tag than the usual gadgets, such as a charity donation. Many people are already doing it. In 2004, for instance, Oxfam sold more than 30,000 Goat Gifts as Christmas presents – a card (and goat picture) for the 30,000 recipients and a potentially life-saving injection of aid to the developing world.

As before, when I began to write about our car use, I can feel myself slipping off down the road of open-toed ranting, calling for rebellion against capitalist values and the immediate incarceration of all shopping channel presenters. Turning our lifestyles back to the dark ages doesn't appeal though. I like TV (well, some of it) and I like having a video recorder and a computer (what a naughty consumer I am), but by avoiding all those extra gadgets we can have these things and still make meaningful reductions in our energy use. Christmas, and in particular the 'consume till we bust' side of it is in for some stick throughout this book – if I'm not destined to be Jacob Marley, then Ebenezer Scrooge certainly looks on the cards. I love Christmas, and I really do dream each year of a white one. It's hard to beat isn't it? The surge of excitement as you wake up on Christmas morning and know from the yellowish tinge to the light and the muffled sounds outside that it has snowed. Keep on with all those gadgets though, and we can all forget about white Christmases and just get used to wet ones.

If you're at home, listen for a minute. Quiet? Listen very carefully and you'll probably just about be able to hear a hum. I can hear it right now. Over on the stereo a little red light gives the game away – standby power. The waste of energy in our households through things being left on standby is a quietly growing monster. We see those little red lights on our TVs, VCRs, stereos and set-top boxes so often that it's all too easy to forget that they are slowly eating up energy (I've just turned the stereo off at the socket). Manufacturers must carry a good amount of the blame for this problem, for making it appear that things are turned off when in fact they are on standby, and for not providing easy ways of switching things off completely, other than pulling them out of the wall.

The spectre of standby power has been with us for many years now, but it is the booming market in household electrical appliances which has sent this source of household emissions soaring. Computers are a prime example. They can take over a minute to start up and so are often left on all the time, the words 'All work and no play makes Jack a dull boy' slowly turning against a backdrop of stars. Screensavers are not energy savers. Switch the monitor off while you're away and you'll halve the energy use of the PC. For always-on PCs the energy saving options can make quite difference. The sleep mode can cut greenhouse emissions by 80%.

Our complacency about background energy use has reached a stage where plug-in air fresheners are now bought in their millions. Together, those quietly humming appliances are eating up over 10% of the electricity used in most homes, just to sit in standby mode. For the average house, three-quarters of a tonne of greenhouse gas is emitted every year by this route. On a national scale the wastage of energy and emission of greenhouse gas from standby power is simply embarrassing. In Australia, standby power is responsible for over 5 million tonnes of greenhouse gas every year and in the USA the figure is closer to 30 million tonnes. All just to keep those red lights blinking.

The third biggest slice from our home-made emissions pie, and one we've touched on already, is water heating. It's not desperately exciting, but at up to 2 tonnes of greenhouse gas per household each year, it's too important to ignore. As with home heating, the big climate-saving actions with water heating are installing good insulation and an efficient boiler. Lagging the hot water pipes can cut the energy-related greenhouse emissions from your water heating by 120 kg a year. Fitting one of those puffer jackets to your hot water tank can cut its energy wastage by three-quarters and save up to half a tonne of greenhouse gas each year.

There's room again here for cuts in energy use through changes in our behaviour and, in particular, in our use of hot water. Visit your own government's Department of Energy (or equivalent) website and somewhere there'll be a page berating you for leaving the tap running while you brush your teeth, using too much hot water for washing up, and having too many baths. For every 15 litres of hot water coming from an electric heater about a kilogram of greenhouse gas is emitted – just fixing a dripping hot water tap can save more than 100 kg of greenhouse gas a year.

What actually powers all the space and water heating, appliances and gadgets in our homes plays a crucial role in determining how much they help drive global warming. Gas heating, for instance, produces about two-thirds less greenhouse emissions than standard electric heating. The electricity supplied to our homes tends to carry with it a rather hefty greenhouse penalty because of the leading role that coal-burning power stations play in its generation. If you're lucky, you'll be able to switch your energy supplier to one providing renewable energy, such as that from wind farms, and so slash the amount of greenhouse gas released for every unit of electricity you use by over 90%.

While we're on the subject of what provides the energy in our homes, and the benefits of using renewable energy, indulge me for a moment while I go a bit woolly-jumpered and look at home power generation. It may conjure up images of bearded woodsmen whittling clothes pegs on their solar panelled front porches, but the generation of renewable energy at home is more widespread than you might think. As well as being climate-friendly, it is less wasteful. It avoids the long-distance transmission of electricity from power stations, provides more energy security in times of high demand or blackouts, and can be cheaper. In terms of emissions, the level of cuts you achieve depends on the scale of action you take. Small solar panels, for example, can be used to charge batteries, provide garden lighting, or top up energy use in the home.

For those homeowners keen to generate a large proportion of their own electricity, an expanse of photovoltaic (PV) cells is the usual option. In sunny climates these can fulfil all the household's energy requirements for much of the year. Even at higher latitudes you can expect an average installation to provide between a third and half of your annual energy use. Most systems allow you to sell any excess power back to the grid, so making you a bit of money and allowing others to make use of your renewable energy.

Such solar systems tend to be expensive, even with government grants to help soften the blow. For a less than sun-drenched area like my own it would take around twenty years for the energy produced by a big set of panels to offset the initial costs of installing it – twice the guaranteed lifespan of the panels. Even leaving the money aspect aside, the energy which goes into producing these solar panels in the first place can be substantial, and take off much of their climate-friendly gloss. For sunny Scotland, I'm looking at between 8 and 12 years of running my very expensive panels before they've produced enough energy to cancel out the energy that went into making them.

The technology, though, is moving on apace, and in large areas of the USA, Australia and elsewhere it is a very real alternative to the greenhouse gas-heavy electricity currently coming down the wires. It's worth mentioning the cheaper, less well-known cousin of solar power generation: solar water heating. At a fraction of the price of the electricity-producing PV cells, and requiring much less energy to manufacture, these water-heating systems – essentially a network of water-filled tubes on your roof – can offset their embodied energy tag and begin saving energy within just 6 months of their installation. In temperate climes, such solar water heaters generally serve to take the edge off the coldness of the water before the heating boiler has to do its work, with a greenhouse gas saving of up to 1 tonne a year. In the sun-kissed suburbs of Sydney, though, such systems can meet the bulk of hot water demands and mean greenhouse gas reductions of more then two tonnes a year.

Wind and water power can also be harnessed to produce some or all of a home's energy needs through small scale turbines and a suitable location. Other options include biomass boilers which use wood chips for home and water heating, and ground source heat pumps which collect heat from the soil around a building and transfer it inside.

Enough, for now, of these home energy and heat-generating technologies. Let's get back to the sharp end of the problem – where we use, and lose, all that energy, renewably generated or not, in our homes. We've already looked at the big three: space heating, appliances and heating our water, so now let's turn to the simplest climate-saving action of all: changing a light bulb.

I can't even begin to write about home lighting without thinking of my father. When my brothers and I were growing up in the 1970s, hardly an evening passed without Dad coming in from work and getting cross at us for the house being lit up like 'Blackpool Illuminations'. This of course was long before the greenhouse effect was making the headlines, but our dad, never a man happy with waste, hated such profligate use of lights (and the resulting electricity bill).

Over the past few decades our lights have got more and more efficient and at the same time we've been using more and more of them. Overall, home lighting is hoovering up even more electricity than it was when my brothers and I were busy making a DIY Blackpool in our small corner of Lincolnshire.

In the top tips books on individual action, climate-friendly lighting always gets a big push. It's obvious why: it's so easy to do. This isn't about our abandoning the car for a bike, spend-

Figure 8 The evolution of Edison.

ing thousands on solar panels, or even reading the labels of the food we buy. Low-energy bulbs are cheap, easy to get hold of, will save us money and will cut our emissions. Why isn't everyone using them already? Goodness knows. For an average house with, say, 12 lights, replacing old light bulbs with low-energy bulbs can cut almost a tonne off the annual household greenhouse emissions – a big result for standing on a chair a few times. Next time you're in that big blue shop of Swedish origin, and before you get side-tracked into 'Beowulf' drinks coasters or 'Hagrid' shelving units, stock up on some low-energy bulbs.

After lighting, the final sliver of the emissions pie is, rather fittingly, cooking. Over a year, using gas for cooking, rather than electricity, will save up to a quarter of a tonne of emissions. Other strategies, such as always putting lids on simmering pots – this uses a third less energy – and filling kettles only as full as required, can make small but effective dents in a household's power drain. If everyone in my home nation of tea drinkers stopped overfilling kettles the energy saved could power more than two-thirds of all UK street lighting.

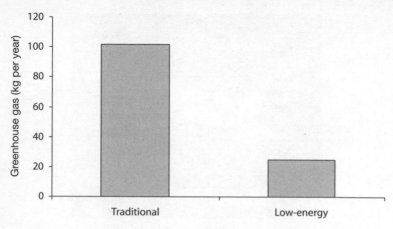

Figure 9 The climate benefits of a low-energy bulb.

We've talked then, about the main areas of our home life that eat through our annual energy budgets and so generate greenhouse gas. Clearly, we can do a lot to make our home life more climate-friendly (Table 2). But there's another big emitter of greenhouse gas in our homes, and more specifically in our kitchens, which uses huge amounts of energy indirectly and whose contribution to global warming is growing fast: our food.

Table 2 At-a-glance ways to cut greenhouse gas emissions from your home.

	Heating habits	Efficient appliances	No standby power	Efficient lighting	Better insulation
Greenhouse gas savings	🌍 Up to 30%	🌍 10–20%	🌍 5–10%	🌍 5–10%	🌍 Up to 40%

4

flying strawberries

It's easy to see why wasting electricity in the home or driving a big car is bad for the climate. Though the carbon dioxide released is invisible, we know that doing these things burns fossil fuel and that, by doing them less, we can cut emissions. Things get less obvious when the emissions arising from our lifestyles are several steps removed from us. The whole embodied energy problem is a case in point. Who would have thought that the innocuous looking gnome in next door's garden carries a climate tag? Sitting there, fishing rod in hand, slowly growing a covering of lichen, its climate impact is hidden in the energy used to mine its materials, to manufacture its cement body and its paint, and finally to transport it from World of Gnomes to your neighbour's garden. Unless you are willing to spend an awfully long time reading some of the driest research papers in Christendom on 'The life cycle analysis of gnomes and their rods' every time you want to buy something then your climate-aware choices boil down to simply buying less stuff.

There is something, though, that we can't just stop buying, and which can carry much more of a hidden climate tag than our bearded fishing friend: our food.

As with the bricks in our walls, the appliances in our kitchen, or the ornaments in our garden, the chain of events which gets a pre-packed cheese and ham bagel onto the supermarket shelf can use a lot of energy. Like a rather tasty snowball rolling downhill toward us, our lunchtime bagel builds up ever more embodied energy, and so more greenhouse gas emissions, on its journey from field to food aisle. Before we expose the bagel for the climate pariah it is, before we get into the nitty-gritty of food miles and belching cows (you'll see), what about our two-way climate change street? If we're going to address the climate impact of our food, what about the impact of climate change on what we eat?

Just as northern hemisphere homes are gradually moving south climate-wise, so are farms. Over the next 20 years many farmers will be able to grow more crops previously the pre-serve of lower latitudes. The flip side of this of course is that the crops they are used to growing might not fare well in such warmer climes. All that extra carbon dioxide we're pumping into the atmosphere means that some crops will grow faster and bigger – so bumping up yields. For some farmers this will be of little comfort after their entire harvest has been destroyed by the more intense droughts, floods and storms we can expect along with the carbon dioxide-accelerated growth rates. Even this so-called carbon dioxide fertilisation effect is not necessarily a good thing. It can make weeds grow faster still, so lowering yields or requiring big increases in the use of herbicides. Overall, some crops will do better, some much worse. For staples like wheat, rice, maize and soybeans, the 2080s could see yields fall by up to 50%, global price hikes of up to 45%, and a 50% increase in the number of people at risk from hunger as a result.

The wine industry in California is worth $4 billion each year, and produces some cracking wines. As summers get hotter and drier over the 21st century, the grape harvest is likely to shrivel and the wine industry with it. Likewise, Californian dairy farming will probably suffer as more intense heatwaves and droughts mean poorer grazing, increased stress to the animals and an overall drop in production of up to 20% by 2100.

More frequent droughts, storms and flooding will disrupt food imports and exports, with increases in pests and diseases further exacerbating these problems. Climate change is going to have far-reaching effects on food in developed countries, but it is food production and supply in the developing world – so often the stalking ground of famine – that face the biggest threat.

In Africa in particular, the outlook for food security is grim. Already nearly 200 million people here are classed as under-nourished, and much farming is on the margins of what the

climate will allow. As the century unfolds, the same intensifica-
tion of droughts, floods and storms that threatens Californian
wine growers could cost the lives of millions of Africans. Globally, we can expect around 80 million extra people to be at risk
of famine due to climate change by 2080, up to 65 million of
these will be in Africa. That's the equivalent of the entire population of Great Britain facing starvation as a direct consequence of
the fossil fuels we are now burning. The very food we eat today
may starve millions in Africa, Asia and South America tomorrow.

And so to the food in our cupboards, fridges and supermarkets: how exactly does it pick up such a big climate tag and
how can we avoid this hidden strip of the grocery bill? It all
starts down on the farm. Agriculture is to blame for almost half
of all human-related methane emissions and three-quarters of
our nitrous oxide emissions.

Take something as innocuous as a wheat field – once a wood-
land, drawing in and trapping carbon dioxide. Each year, after
the harvest, the soil is ploughed. This exposes deeper, carbon-
rich soil to the air and leads to a flush of carbon dioxide into the
atmosphere. To get the next lot of wheat growing fast, farmers
heap on tonnes of nitrogen fertiliser, increasing emissions of
nitrous oxide. The same is true of just about every crop – add
nitrogen fertiliser and you get nitrous oxide.

Then there's methane. The bugs that produce the gas like all
things wet. If you've ever plunged your booted foot into a
waterlogged bog and seen bubbles rushing up from the ooze,
usually along with a smell of rotting vegetation and the creeping realisation that your boots aren't waterproof, then you've
seen methane escaping. As with boot-eating bogs, the satu-
rated soils of paddy fields house countless billions of methane-
producing bacteria. These pump out around 60 million tonnes
of the stuff a year.

As the consumers of these crops there's little we can do about
such emissions (apart from avoiding wastage and growing our

own). Governments around the world are already trying to limit how much nitrogen fertiliser goes on fields, not least to stop it getting into our drinking water when it gets washed off. Rice farmers too are increasingly encouraged to let flooded soils dry out a bit more often and so keep a lid on all that bubbling methane. Where food-related greenhouse emissions start to get really big, and where we do have some choice, is with meat.

Imagine your average cow (Figure 10). As well as eating lots of grain grown in the fertiliser-soaked, nitrous oxide-emitting fields, she also emits buckets of methane. As she trundles around the field this cow and all her bovine friends belch out methane, day in day out. In just one day your typical Daisy burps around 200 litres of methane into the atmosphere. Where you've got a lot of cows this adds up to a great deal of methane. In Australia for example, the climate impact of

Figure 10 Bucolic belching.

carbon dioxide is in danger of being overtaken by that of live-stock-burped methane – at a massive 3 million tonnes a year. Something similar is going on in New Zealand, in large part due to its 50 million or so sheep. Even in the USA, where energy use and transport dominate greenhouse emissions, methane from cattle tops 5 million tonnes a year. Cattle, the biggest of our methane-emitting farmyard friends, are not the only ones burping their way to a warmer world. Sheep weigh in with about 30 litres of methane per day each, while pigs emit about 8 litres.

Yes, some humans can also release large volumes of methane (and this time it's not belching we need worry about). Thankfully, the condition is quite rare. Whereas most of us let out just a few hundred millilitres in the privacy of our own bathrooms, some, probably quite lonely, souls emit as much as 3 litres of methane per day.

Meat, then, has often built up a big climate burden when the time comes for that one-way trip to the abattoir, thanks to the energy and emissions associated with feed and methane. Every beef steak results in around 15 times its own weight of greenhouse gas even before you add in transport. In the Netherlands, a nation very fond of a bit of animal for breakfast, lunch and tea, meat contributes almost a third of all food-related emissions. Dairy products, incidentally, with their obvious requirement for more belching cows and grain, make up nearly a quarter of Dutch food-related emissions.

Even without the heavy greenhouse gas tag, meat's statistics are pretty sobering. Currently over 800 million people are either undernourished or lack a secure food supply; water is critical in getting these people the food they need. Much of that precious water goes into growing food for livestock. Producing a kilogramme of beef uses 15 cubic metres of water; a kilo of cereals needs less than 3 cubic metres. Meat uses up to seven times more land than crops of the same food-value.

Since the Second World War the amount of meat each of us in the West eats has soared to 100 kg a year. That's a steak per person per day.

I enjoy meat. No barbecue in my street would be the same without some charcoal sausages and my neighbour's wonderful spicy kebab. Where things have gone wrong is in the sheer volume of the stuff we now get through. Meat used to be a treat, something which cost quite a lot and wasn't always available. With intensive production methods, farmers and supermarkets have pushed the price of meat down to a level where a whole chicken costs less than a high street coffee. Now we feel that a meal isn't a meal unless a large chunk of animal flesh is involved. These low prices hide big social, environmental and medical costs, as flagged so articulately in Felicity Lawrence's recent book *Not on the Label*.

The fact remains that meat, no matter how well looked after, is bad for the climate – another moral fly in your oxtail soup. Cutting down on the amounts we chomp through can make a difference on several levels. If you long ago shunned meat on the grounds of your health or animal rights, you can congratulate yourself that at the same time you've also been cutting swathes through your emissions. If, like me, your mouth still waters at the thought of rare steak, then at least cut back a bit, with the added bonus that you'll be healthier as a result. Just two fewer beef burgers or steaks a month can slash your annual greenhouse gas output by a third of a tonne.

The research on food, its embodied energy and its contribution to global warming, is nothing if not thorough. I have before me papers listing the embodied energy of everything from unsalted nuts to block margarine, from wholemeal bread to fortified wine. A particularly diligent group in Sweden have calculated the energy which has gone into not just pork, but fresh pork, frozen pork, pork from Sweden, pork from Europe, pork stew and pork sausages.

For the whole life-cycle of a meal, from field to plate, the differences in the greenhouse gas emissions for vegetables compared to meat are stark. So let's take a trip to Sweden, land of saunas, home furnishing and, yes, ABBA. After a heady night of outrageously priced beer and a morning of groaning and sipping water, we make our way down to the hotel's restaurant for the buffet lunch.

On the menu today are five appetising options: roast pork, boiled carrots, fresh tomatoes, fried potatoes and beef stew. If each of these trays of food had its climate tag they'd be labelled as follows: 50 g of carbon dioxide for a serving of carrots, 330 g for some of those shiny tomatoes, a mere 17 g for those great smelling potatoes, but an appetite-quelling 610 g for a few slices of pork and a massive 1,500 g of greenhouse gas for a couple of scoops of beef stew.

The tomatoes, grown in Sweden but in heated greenhouses, have clocked up a big climate tag through the oil-powered heating systems used to keep the glasshouses balmy even while the snow is settling outside. Storage, and in particular refrigerated storage, eats up energy too. For the carrots, 60% of their total emissions arise through keeping them chilled until the supermarkets want them. And then there's the extra climate whammy of food transport.

Yes, some foods have already packed a big climate punch by the time they pass through the farm gates. For others the contribution to global warming really starts to build up as they set off on what is likely to be a very long journey, crossing borders and even continents.

Pick half a dozen food items at random from your cupboard or fridge and you'll probably have before you food which together has travelled thousands of kilometres – now dubbed food miles – to your kitchen. The six rather tasty fridge items in Figure 11 would together travel over 40,000 km in getting to my fridge in Scotland, were I not such a seasonal, local pro-

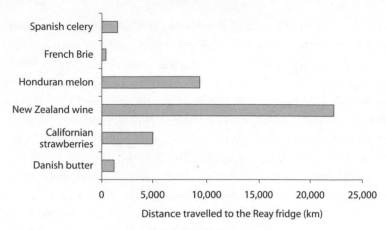

Distance travelled to the Reay fridge (km)

Figure 11 Around the world in a fridge.

duce/grow your own evangelist. This long-distance transport of grub means the burning of a lot of fossil fuel and so the emission of a lot of greenhouse gas.

Our Swedish buffet was relatively food-mile-lite, much of it having been produced within Sweden. What about those luxury or out-of-season foods which have sailed the seven seas, been trucked the length and breadth of continents, or garnered enough air miles to qualify for a free Harley-Davidson? Their tags are big. A recent study found that one basket of 26 organic items clocked up over 240,000 km and 80 kg of greenhouse gas emissions before being pedalled home by a right-on shopper.

Take your lunch at work: it's rarely anything to write home about, more a fuel stop to get you through the next four hours of keyboard tapping. You pop out to the shop and grab that pre-packaged cheese and ham bagel, a drink and some fruit, paying handsomely for this collection of clear plastic containers. Back at your desk you rip open your bagel, stop to staunch the blood flow from your newly lacerated thumb, and munch down the chilled offering while typing 'packaging injury and litigation' into Google.

Producing the ham and cheese, as we've seen, results in lots of greenhouse emissions. Getting them into our bagel then racks up yet more emissions during processing, refrigerated storage and 1,000 km of travel. Add to these emissions those of the bread and butter surrounding them and you have a bagel with an overall climate tag of almost half a kilogram of greenhouse gas. That's one heavy snack, even before we start on all the plastic packaging, of which more later.

Feeling health-conscious you also bought a bunch of grapes and a bottle of mineral water. Now things take a real turn for the worse. The bottle of spring water, with its label showing a babbling brook in Canada, has travelled nearly 6,000 km by boat and truck to get to the supermarket, where it has since sat in a refrigerator. This one bottle, which, let's face it, contains an expensive version of what we get out of the taps, has clocked up nearly 300 g of greenhouse gas – truly heavy water.

The grapes are the killer. So far this lunchtime our food and drink has crossed oceans and international borders using boat, truck and van. The grapes bring us to air freight. It is preposterous that while worrying about exhausted oil reserves and the environmental damage caused by the rapid growth in air travel, we continue to fly tonnes of exotic fruit and vegetables around the world.

Your 200 g bunch of grapes has clocked up over 10,000 km on its way to your desk, most of that by jet. Its transport has added one and a half kilograms of greenhouse gas – six times its own weight in emissions and equivalent to leaving a climate-unfriendly light bulb on all weekend. If the mineral water is like a lead milkshake as far as global warming is concerned, then these grapes should sit in your stomach like a bag of ball bearings.

Air transport of food has reached ridiculous proportions. Every new and exotic ingredient used by the celebrity chef of the moment sparks a surge in demand for fresh tiger prawns or papaya, all flown in that same day. Felicity Lawrence tracked the life and times of a hand-tied bunch of chives which says it all. Though the chives started off in the UK and were destined to be bought and eaten in the UK, the little bunches were first flown to Kenya for the 'hand tying' before being flown straight back, thereby clocking up about 14,000 km and over a kilogram of greenhouse gas for every 20 g bunch.

All over the developed world the story is the same. A bunch of grapes at a Chicago market in 1972 would have travelled about 2,500 km; by 1989 Chile had become a major source of grapes and the average distance each bunch was travelling had almost doubled.

It would be daft to say that we should all consume *only* locally produced or home-grown food and spend our winters smeared in goose fat and sewn into our long-johns, eking out the last few shrivelled apples from the autumn. But simply sourcing some food locally can massively cut emissions. Apples, shrivelled or otherwise, are a good example. In the UK the supermarket shelves are dominated by imports; even during the British apple harvest most of the varieties on show are imported from New Zealand, Australia or South Africa. Buying local apples, instead of travellers from the Southern Hemisphere, can cut the related greenhouse gas by nearly 90%.

The icing on our well-travelled climate cake is shopping trips. The number of shopping journeys we make and their distance has grown in recent years – over 200 shopping trips a year on average. Strolling down the high street to the butchers, bakers or grocers clearly leads to lower emissions than getting in the car. Increasingly though, high street food shopping isn't an option – few of us have local stores anymore. Instead we have out of town malls like aircraft hangars with many kilometres of roaring tarmac between us and their hectares of parking lot. These barns, purveying everything from Bleu d'Auvergne to porcelain statuettes of Kermit the Frog, should mean we take fewer trips. They don't.

A 25 km shopping trip in the car equals emissions of 5 kg. Divided between all the items in your trolley this trip adds just 125 grams to the climate tag of each. It's the extra 'nip out' for a bag of mixed nuts that can dump a whole 5 kg of greenhouse gas on your nutty snack. Cutting out these extra trips by always making a list and resisting the temptation to grab the car keys every time you see an advert for a new flavour of nachos is a start.

Another, very 21st century, option harks back to the days of delivery boys on wobbly bikes: Internet shopping. One truck can deliver the weekly food shopping of dozens of families, preventing dozens of trips by car and reducing greenhouse gas emissions by between 20% and 90% (depending on how close together the drop-offs are).

Climate-aware food shopping is a prime example of the clash between 'doing your bit' and real life in the developed world. Food shopping is about keeping the family fed for the coming week on a limited budget, and getting out of the shop before your last strand of good humour snaps and the kids take to beating each other with bags of toilet tissue.

Even those unhurried 'ethical' shoppers with baskets full of expensive organic products and an air of smugness (they undoubtedly model for energy-saving pamphlets) can end up laden with greenhouse gas. Indeed, it's these shoppers who are most likely to buy hand-tied chives to serve at that night's dinner party with organic baby sweetcorn from South Africa, lamb chops from New Zealand and wine from Australia, while wringing their hands over the recent flooding.

We're bombarded with messages about our food all the time: eat less fat, eat less sugar, buy organic, buy fair trade. Adding to this list 'buy low food miles' is vital but difficult in practice. Labels tell you little about climate impact and supermarkets' range of local produce is often limited. If we knew which foods were global warming gremlins and which weren't, if labels carried the number of food miles for instance, we could make informed choices. This would put pressure on the supermarkets to sell more local produce.

One suggestion is ClimateWatchers, a greenhouse gas-aware version of WeightWatchers. The idea is that people will go to meetings with till receipts and be told how much weight (of greenhouse gas) they've gained or lost. It conjures images of 'Climate Slimmer of the Month' – a bearded man who has lived solely on nettle tea and boiled limpets for the past four weeks – but the idea is sound. Until branches of ClimateWatchers spring up in church halls we'll have to rely on Country of Origin stickers to help us eschew globe-trotting food and the threat to our climate it represents.

In the developed world the average person emits an extra tonne of greenhouse gas each year through the food they eat. A weekly trolley load of globe-trotting supermarket produce can push that to over 4 tonnes a year. With a climate-aware diet on the other hand – local produce, no flying food – annual food-related emissions can shrink to around a third of a tonne: a 90% cut (Table 3). Switching to a life of wild berry eating, interspersed with the occasional grub or roasted squirrel, isn't necessary.

Table 3 Potential cuts in food-related greenhouse gas emissions.

	Less meat and dairy	Fewer food miles	Fewer shopping trips	Food delivery	Home grown
Greenhouse gas savings	Up to 30%	Up to 90%	5–10%	5–10%	Up to 100%

↔

Assuming we've cut back on steaks and sneered at the exotic fruit section for most of the year, there are surges in consumption, of food and everything else, that still lie in wait. Weeks when the idea of keeping an eye on the food miles, the amount of meat in the shopping basket, and our own waistlines tends to go out of the window: celebrations like Christmas and Thanksgiving.

Yes, it's Scrooge time. We're talking about food, and these festivals for most of us revolve around meals of gigantic proportions, so let's get out the humbugs.

It's Christmas Eve and the smell of pine needles is in the air, radio stations are stuck in a repeating loop of Bing Crosby, and the arguments about whose relatives should visit or be visited have been settled. For the Carbones, Christmas is a traditional affair. Grandma Carbone comes to stay and the kids work themselves up into a frenzy of over-excitement about the presents they hope to get. The tree is already standing in the front window, bedecked in bright fairy lights, miniature elves and slowly rotating baubles. The glitziness of the tree is nothing compared to the decorations outside. John Carbone, determined not be outdone by the neighbours, has really gone to town this year. The front hedge winks on and off under the power of a hundred bright white bulbs. Along the eaves glowing white 'icicles' are picked out by ripples of light travelling backwards and forwards in never-ending waves. On the roof a brightly lit Santa Claus waves to all who pass by, while resting against his gift-packed sleigh and four reindeer, themselves picked out in hundreds more brightly coloured bulbs. In the front garden the Carbones' one tree hasn't escaped: every branch is covered in closely packed fairy lights that appear and disappear at apparently random intervals.

John and Kate are scrumming down with, seemingly, the rest of humanity at the local mall. Having done half a dozen circuits of the car park and finally found a space, they trudge the half kilometre to the stores and hit the blazing lights and the wall of jingly sound that is the Christmas shopping experience. Their aim is to get the last few presents for the boys and buy enough food to last through to Boxing Day, all without losing the plot and attacking the nearest group of handbell ringers.

The presents for the boys are, without exception, electronic gadgets, so one shop meets all their needs. For George a new cell phone is a must, as is a new games console. For Henry a new laptop is his main present, with another new mobile to boot. John and Kate also take the opportunity to buy a big, new plasma screen TV – something John has been lusting after for several months.

Having run the gauntlet of the electrical goods store, tried 23 different cell phones, and plumped for 'Dictator: Crush and Control' as the game to go with George's new console, it's off to the food hall to do battle with the other 200 people over that last tub of fresh cranberries. Grabbing a trolley each, and consulting the long shopping list, they launch into the melée and begin to collect the food planned to satisfy the family over the coming days of over-indulgence.

Three hours later, and with the mad shopping experience over (at least for a couple of days), the Carbones make their way home to get the food put away and the presents wrapped before the boys' prying eyes seek them out. Grandma Carbone arrives just before supper-time and, as the Christmas Eve excitement builds, the Carbones sit down to the first of many blow-out meals.

Christmas Day dawns warm and wet, much to the disappointment of the whole family, who were hoping against hope for a white Christmas. It's been 20 years now since it snowed on Christmas Day. The sledges in the shed might get an outing come January, though even that's not a sure thing. With Christmas carols on the stereo competing with *It's a Wonderful Life* on the TV, everyone gathers for the gift opening. Paper is torn, novelty ties are put on, batteries are hunted down for new gadgets and dust jackets of new books are read. After a long session of present opening, the table is set with the best cutlery and glassware and it's 'all hands on deck' in the kitchen as a turkey the size of a small ostrich comes out of the oven, followed by trays of ham, stuffings, and mashed potato. Lashings of hard-won cranberry jelly are scooped into dishes and pints of sauces come off the hob. In the dining room the diners take up their positions for the feast ahead and, as the platters come through, the table groans under their weight. Just as it seems the table has reached its limit, out comes the champagne, the wine, the beer and the cola. Glasses are filled and raised all around and the feast begins. Some considerable time later, after the final spoonful of mash has been forced down and that slice of cherry pie has accounted for the last notch on John's belt, it's time to load the dishwasher and slump down in front of the new TV for the Christmas Day movie. With luck, when the time comes for another few plates of cold meat and pickles everyone's stomachs will have shrunk slightly and the bloated pain they are all suffering will have subsided. After some colourful swearing at useless instructions and missing connectors from John, George quickly takes over and gets the new plasma TV set up and running. The surround sound is soon joined by the combined snoring of Grandma and John.

Later, as the house lights dim due to John flicking the switch for the outdoor illuminations, the Carbones can reflect on another Christmas Day done. After another raid on the bowls of nuts and sweets, a new DVD, and an argument about who will go to the sales the next day and what food needs to be bought, it's off to bed for George and Henry, and a glass of eggnog all round.

The Carbones' Christmas Day food consumption is average for the developed world. Together, the bowls of nuts, boxes of sweets, and the stocking satsumas, the turkey and trimmings, lashings of wine and spirits, and chocolate, add up to emissions of well over 25 kg – double those of a normal day. For the Christmas holiday as a whole each household belches out about a quarter of tonne of greenhouse gas just from food. And it doesn't stop there.

There's the extra power use, in particular those Christmas lights. For the Carbone household, the hundreds of bulbs, not to mention the waving Santa, burn around 200 kilowatt hours of electricity during the Christmas period, leading to an extra 120 kg of greenhouse gas going up into the atmosphere and helping to ensure that next Christmas won't be a white one either. Across the USA, festive lights fritter away 2 billion kilowatt hours every year: enough energy to power 200,000 homes all year round.

A more lasting drain on electricity, and belcher of greenhouse gases, comes in the form of all those presents, each with its embodied energy tag. Most homes on a Christmas afternoon suffer from a new rash of blinking and buzzing gadgets, some eating their way through their first and last set of batteries before being relegated to a dusty cupboard. The more tenacious electronic devices soon take their place in the standby power hum and the three-quarters of a tonne of related greenhouse gas emissions.

Finally, there's the most conspicuous side of our rapacious Christmas consumption: rubbish. After the big day and all that present opening, most of us feel we need to drag ourselves out for a bit of fresh air to try to digest enough food in our stuffed stomachs to make room for the next round. The bins outside every house tell the story of just how much consumption has been going on inside. No bin is big enough for the onslaught of wrapping paper and packaging, discarded food and bottles. As the garbage collectors are usually at home creating their own mounds of trash, Christmas and Boxing Day are the few days each year when you really get an idea of how much stuff every house is throwing out. Street corners rapidly fill with bulging plastic sacks, Santa-covered paper peeking over their tops or blowing down the street. Another week and the Christmas trees start to appear, sad browning skeletons of their former bushy selves too big to get in the bin so leaning dejectedly against the wall until the next collection. All this waste means a lot of truck collections and some very rapid filling of landfill sites. Each Christmas the UK throws out an extra two and a quarter million tonnes of waste food, wrapping paper and packaging, plus about 6 million Christmas trees. In the USA the numbers are even more boggling: 33 million Christmas trees are dumped each New Year. This Yuletide bonanza for the landfill bugs means an extra 200 kg of greenhouse emissions per household.

An average Western Christmas – the food and the presents, the lighting and the waste – adds up to a badly wrapped present to the atmosphere of over half a tonne of greenhouse gas. The most direct way to reduce these emissions is to cut back on the amount of stuff consumed in the first place and direct those tonnes of wrapping paper, silly hats and old sprouts somewhere other than a landfill site. Which brings us neatly to our backyards – a potentially big source of planet-warming emissions, and a great place to slash them.

5

in my backyard

Our backyards are increasingly extensions of our homes. Turn on daytime TV and there's certain to be some green-fingered presenter telling you how to make the garden an extra room of the house. Now, in even the most frigid corners of Europe and North America, you can find families out on the decking having a barbecue, year-round. It may get cold (though less so these days), but there's always the forest of patio heaters to keep us warm – the last word in climate control. With this deepening love affair with outdoor living has come a whole new industry: from those winsome TV gardeners, to garden centres selling everything from electronic chimes playing Slim Whitman hits to grass that glows when you walk on it. Yes, we love the great outdoors.

Climate change is set to spoil the party. The TV garden shows of 2050 will be less 'Hanging baskets and the need for threes' and more 'My lawn's dead and the kids have got ticks'. What's in store at the bottom of the garden? It's not *The Good Life*, that's for sure.

As you tentatively stick your nose out of the door, you might well encounter the first impact of climate change in the very air: the smell of smoke. With the intense summer droughts forecast to result from global warming, forest and range land fires are also predicted to increase. Hot-weather smoke isn't just an annoyance if you've put your washing out to dry, it is bad for your health. When several major forest fires broke out in south-east Asia in 1997 their combined smoke darkened the skies for weeks and led to a tripling of respiratory disease cases and a drop in lung function in schoolchildren. The next year, raging fires in Florida doubled the number of people visiting the ER with asthma and bronchitis and upped the admissions due to chest pain by a third.

Studies have also linked baking summer weather to increased radon concentrations (a cause of lung cancer), increased ground-level ozone (causing respiratory problems), and health damage from a host of other air pollutants. Across Europe, more than 300,000 people already die prematurely each year because of air pollution.

If you haven't coughed from the smoke and other air pollutants you'll probably sneeze. Every hay fever sufferer knows that long periods of hot dry weather make going outside at all something which should only be attempted with an inhaler and swimming goggles. Tree pollen is likely to rise along with summer temperatures, and the carbon dioxide fertilisation effect will make weeds grow faster and produce more pollen. Experiments have already shown that a doubling in carbon dioxide concentrations leads to a quadrupling in ragweed pollen.

As we snuffle and hack down the garden path, the health threats continue. A changing will make our gardens home to a host of new pests and pathogens, as well as changing the numbers and behaviour of the ones we have already. In the early 1990s, following a period of exceptionally heavy rainfall in the USA, the mouse population exploded. Along with a rash of mouse dropping-covered cupboards came the first recorded US outbreak of Hantavirus Pulmonary Syndrome (this is one nasty disease: sufferers' lungs fill with fluid and over a third of those who contract it die). Warmer winters will spur other diseases, as will increased flooding. Extreme weather may damage health services and malnourishment will weaken immune systems. All in all a perfect recipe for epidemics. So far, climate change isn't thought to have directly caused the resurgence of any infectious diseases, but chances are that it will.

Topping the list of diseases threatening to spread to a garden near you is malaria. Forty per cent of the world's population are currently at risk from the disease; over two and half million people are now infected and it kills a million each year. A child dies from malaria every 30 seconds. It is on the rise across the globe, because of poor public health, pesticide-resistant mosquitoes, and increased human movement. Climate change is the last thing this deadly mix needs.

Malaria doesn't like it too hot, and where it's already living at really high temperatures – in some areas of Kenya, for example – further rises in temperature could kill it. For most of the world though, the mosquitoes and their malarial payload will lap up the heat, with even small rises massively boosting transmission rates. By 2080 malaria's global reach is expected to have extended to encompass an extra 300 million people. In South America, the range of the engagingly named malaria-carrying mosquito *Anopheles darlingi* is at the moment limited by temperature; climate change may see it extending further south into Argentina, exposing more people to the disease.

Much of the developed world got rid of malaria back in the 1950s and 60s. What we didn't do is get rid of the mosquitoes which carry malaria – something all too obvious on a warm still night when, as you tuck into your climate-friendly barbecued mushroom, you realise that half a dozen of the little beasts are tucking into you. Thanks to scattered reintroductions of malaria by sun-tanned neighbours returning from holiday with tales of great scenery but unexplained cold sweats, the potential for more frequent outbreaks is very real.

Assuming that our health services remain intact in the future we should be able to head off such invasions and limit the damage. More of a worry is when outbreaks occur in countries like Russia, where the medical infrastructure is more fragile and the disease could get out of control.

Of the other diseases more likely to visit our backyards, one of the most dangerous is dengue. Like malaria, dengue is on the rise, with more than half of the world's population now at risk. It too is transmitted by mosquitoes and there have been a growing number of imported cases and outbreaks in the USA and Australia in recent years – it's estimated that travellers introduce around 200 cases to the USA each year. Dengue causes fevers and, in its haemorrhagic form, kills 1 in 20 of those who contract it.

Probably the best-known mosquito-borne nasty on the march is West Nile Virus. This virus was first isolated from a woman in Uganda 1937 and until the 1990s it seemed the Western hemisphere was free of it. Then, as summer turned to autumn in 1999, unexplained cases of encephalitis – inflammation and swelling of the brain – started cropping up on the east coast of the USA. The outbreak centred on New York City and coincided with the sudden deaths of crows and other wild birds. From one of these dead crows a lab in Iowa isolated a virus, later identified as the same one causing the outbreak of encephalitis in humans. West Nile Virus, previously known in Africa and Asia, had arrived in the West. By the time winter came and the mosquitoes were hidden away awaiting spring's warmth West Nile Virus had caused 62 cases of encephalitis, seven of them fatal. In subsequent years the virus rolled out across the USA. By 2003 there were 9,800 confirmed cases in 46 states; 264 died. The true number of Americans infected may be a quarter of a million, as in most instances West Nile Virus causes no symptoms. It is in the elderly or weak that severe illness can ensue. There is no treatment for the disease and no vaccine to prevent it.

Meanwhile, in Southern Europe, leishmaniasis is on the move. Passed on by sandflies, this disease causes sores, fever and weight loss; untreated, it can kill. Other diseases whose ranges could expand include bilharzia (fevers and liver damage caused by parasitic worms); Chagas' disease (passed on by the Assassin Bug and already found in the southern USA); and that old chestnut plague – still absent from its old killing fields of Europe, for now.

Finally, much of the USA and Europe is set to be overrun by ticks. If you own a dog and live in the country you've probably already encountered these tenacious hitchhikers. They sit in the grass waiting for an unsuspecting sheep, deer or Labrador to pass; then they jump aboard and burrow down to the skin for a good meal of blood. Ticks are not averse to a bit of human too and can carry a host of diseases. During their bloody feasting they can pass on Lyme disease, which can cause fevers and fatigue; Tick-borne Encephalitis Virus, which can result in inflammation of the brain; and something called Rocky Mountain Spotted Fever, complete with vomiting and abdominal pain. Already tick numbers have soared in North America and Europe thanks to 20 years of warmer winters. Lyme disease cases in the USA jumped from 491 in 1982 to over 23,000 by 2002.

The advice from the US Department of the Interior is to keep skin covered when outside and to stay indoors around dusk and dawn. Those neighbourhood barbecues clearly have an additional danger to that of being given an undercooked chicken leg. And talking of undercooked chicken legs, food poisoning will also soar with rising temperatures. In the UK the prediction is that climate change will add about 10,000 extra cases of food poisoning each year by 2050 – not just because we'll be eating more dodgy barbecues, but because higher temperatures will mean shorter shelf lives for many consumables. For most, such food poisoning means spending two days within rushing distance of a toilet, but for the frail and elderly the consequences can be fatal.

With rat, fly, mozzy and tick-borne diseases waiting in the fields, parks and streams, and salmonella-garnished burgers crashing our barbecues, and the very air making us wheeze like a forty-a-day smoker, we'd be excused for staggering back indoors for a lie down. But while we can retreat to the air-conditioned cool, our gardens will take the full brunt of global warming. How are those beloved begonias going to cope?

In general, the story is the same as for crops. More carbon dioxide in the atmosphere will make many plants grow faster. Rose fans can expect more buds and earlier blooms. In Europe's milder, shorter winters we'll be able to sow seeds earlier and many of our plants will enjoy longer growing seasons. In the UK, spring already arrives between 2 and 6 days earlier per decade, while at the same time autumn is being pushed back a couple of days. Warmer conditions will alter the types of plant that can successfully grow in our gardens – swapping hostas for poppies and sprouts for kumquats. My Reay Rioja could actually be drinkable. Indeed, large-scale vineyards may spread as far north as Scotland by the middle of this century.

On the downside, soaring summer temperatures and warm damp winters will put an end to the herbaceous border and reduce the rockeries of heather and slow-growing Alpines common to many northern gardens to... well, rocks. Most of us can also resign ourselves to many more hours on hands and knees pulling out the weeds that have lapped up extra helpings of carbon dioxide. Even worse, the lawn may start needing a cut in late autumn or even during winter, assuming it survives that long. With water supplies stretched to breaking point during summer and hosepipe bans in place, the days for many of those precious green squares are numbered. The increasingly desperate battle of the brown patches will be waged across the south of England, home of striped lawns and garden parties. In a country that has always been able to rely on the summer rains and cool days so beloved of lawns, the hot dry summers of the 21st century will change the traditional garden irrevocably. Instead of cream tea on the grass, it'll be chilled water in the shade of the gazebo.

Assuming our plants weather the extended droughts and scorching temperatures, there are yet more unpleasant surprises in store for them. Longer growing seasons will give aphids, mites and thrips – already common garden pests – many more breeding cycles, enabling them to dominate plants from much earlier in the year. Milder winters will also allow larger numbers of these bugs to survive through to the following spring, when they will rampage through the first green shoots of the year. Take the cabbage aphid: for every 1 °C temperature rise they will be able to start their attacks two weeks earlier. For cabbage root fly, an increase of 2 °C will allow them to attack cabbage roots a whole month early (bad news all round for cabbages).

Some infestations currently confined to greenhouses will notice that, actually, it's now rather nice outside and make their escape. Fungus attacks are likely to soar, as the amount of lush vegetation increases, and winters become ever milder. Still more pests will extend their ranges into regions previously too cold. In Europe, termites are tramping north, already cropping up in the south of England. In the USA, trees are falling victim to attackers, such as spruce budworm, previously confined to the warmer south. Overall, our gardens can expect a whole new influx of exotic species, invited or not.

Climate change is coming and none of the standard 'Not In My Back Yard' protests will stop it. Do get out the 'Save Our Neighbourhood' signs and write angry letters to your MP or Senator, but she's probably at home nursing her family through a particularly severe bout of food poisoning, while occasionally sneaking out to break her own hosepipe ban. If you want real action, if the prospect of your garden becoming a parched health risk makes you angry, then it's down to you. In our backyards this action starts exactly where we left our food: in the bin.

There sitting outside the back door is that seemingly bottomless pit for the detritus of Western life. Time was that all that went in the trash was ash from the fire and a few food scraps, but as packaging and consumption have grown so have our bins. As the bins expand we throw more trash in them. Today's average bin will swallow up 230 litres of garbage – that's the equivalent of about three cleaning-obsessed old ladies. It's easy to chuck everything that you don't want into its gaping black maw and forget about it. This is no magic box though. Our trash may disappear from view but it has to go somewhere. Before we look at exactly where all this stuff ends up, let's get out there before the garbage collectors and take a look, like some sweaty paparazzo combing through your week's rubbish for the next big scoop.

An average wheelie bin gathers about 20 kg of trash every week. In other words, over a year, each of us ditches 10 times our own weight. If you were brave enough to push your arm in and pull it all out you'd find something like the breakdown shown in Figure 12.

The biggest part, and most likely the smelliest, is organic waste. This is the remains of last night's pizza, the leftover salad, and the contents of that cup at the back of the fridge which might once have been egg yolks but which now only DNA testing could identify for sure. Across the USA, 25 million tonnes of food are discarded each year. For many with gardens, organic waste also includes the grass clippings from the lawn and what's left of that dead plant bought on a whim from the garden centre. It's estimated that each of us produces between 50 and 125 kg of plant waste every year – nearly 30 million tonnes of yard trimmings in the USA alone.

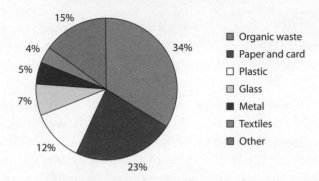

Figure 12 Components of domestic waste in Europe.

All this organic waste, from teabags to banana skins, from apple cores to bindweed, produces that powerful greenhouse gas methane when trucked off to the landfill site. Here about 60% of all trash is compacted down and, when no more can be crammed in, is covered in a layer of soil and left. Down in the dark bacteria get digesting (these are the same bacteria that churn out methane from squelching bogs). Worldwide, landfills emit around 50 million tonnes of methane each year, much of it from the breakdown of kitchen and garden waste. In the USA about half of all landfill methane is now captured and used to power turbines, but the rest bleeds out into the atmosphere.

In short: throwing all those coffee grounds and hedge clippings in the bin is to be avoided. The simple way for a family to starve the methane-producing bugs involves something gardeners have been up to for generations: composting. The average household chucks out around 3.5 kg of food every day, more than two-thirds of which could instead go on a compost heap or into a worm farm (Figure 13), along with the majority of garden waste.

I'm rather a compost nut. The idea that I can throw rubbish from the kitchen and garden atop a heap, give it the occasional fork, and take buckets of crumbly nutritious 'black gold' from the bottom to fertilise the plants a month later never ceases to

Figure 13 The author's composting worms busy making 'black gold'.

get me excited. (In West Lothian you have to get your kicks where you can.) Even better than the compost heap is the worm bin. This converts all the motley cabbage leaves, potato peelings and other organic waste from the kitchen into yet more black gold, and in double-quick time.

It's the ultimate in garden-friendly recycling. Not only do you avoid all the methane emissions, but, after a few weeks, your old carrot tops have become a top-grade soil dressing enabling you to grow even better carrots next year. For every kilogram of potato peelings, teabags or grass clippings that you compost rather than send off to landfill, you'll stop around twice that in greenhouse gas from going into the atmosphere. Over a year, the average household can shave almost a tonne off its emissions in this way. Not bad work for a few worms.

For the garden-less or worm-averse, centralised composting schemes are springing up in large towns and cities. By collecting your organic waste and then piling it onto giant-sized community compost heaps, rather than trucking it out for burial, local authorities end up with tonnes of free compost which they can then sell to garden centres. These huge heaps of collected organic waste are turned regularly to keep up the supply of oxygen and keep a check on the stinkier, methane-producing sections of microbial society. Such a scheme in Albuquerque composts nearly 10,000 tonnes of grass clippings and the like each year, preventing over 4,500 tonnes of greenhouse gas from being emitted.

Back to our eviscerated bin. Paper is the next biggest component in the pile of trash tipped across our pavement. Those spaghetti sauce-covered phone bills and till receipts may be the stuff of dreams for a bin-rifling paparazzo, but they also represent a large chunk of our waste's contribution to global warming. An average family throws out 8.5 kg of paper and card every week, half of this as newspapers and magazines. In the USA, around 85 million tonnes of paper and cardboard are chucked each year. More than 80% ends up wrapped around the TV dinner scraps in the dank depths of landfills. It's not quite as sumptuous a meal for methane-belching bacteria as banana skins with a teabag compote, but it is a welcome accompaniment – an appetiser before the main course, so to speak.

Recycling is an increasingly available alternative here. Not having to chop down and then process trees, with all the associated waste this creates, saves an awful lot of energy. In Australia, the greenhouse emissions arising from making paper total more than 12 million tonnes. Recycled paper takes a third to two-thirds less energy to make than the same amount of virgin paper. So every tonne of paper or card we recycle prevents up to two and a half tonnes of emissions. Every Sunday paper you put out to recycle, rather than heave into the bin, can therefore stop 2.5 kg of greenhouse gas – 150 kg's worth in a year's worth of Sundays.

The motto on every council leaflet about recycling and waste awareness website is 'Reduce, Reuse, Recycle'. It's estimated that for every tonne of bin-bound stuff another five tonnes of waste was produced during its manufacture, and a massive 20 tonnes of waste at the point of extraction. This takes into account all the resources dissipated, say, at the mine where the iron ore that went into making a baked bean can was extracted, or by the logging company that sliced its way through a forest to make the paper label on the tin.

Route one to reducing the amount of stuff thrown out in the first place is less packaging. It seems everything is wrapped and rewrapped as though it needs to survive re-entry from space. Some shops now sell single bananas in a heat-sealed moulded plastic sheath. Yes, a fruit you'd have thought already comes with a rather handy natural wrapper. Then there are the burger bars: you pay an hour's salary for more fat in one meal than is recommended for a week and, along with this heart-stopping culinary delight comes an avalanche of polystyrene containers, napkins, wrappers, bags, cups, lids and, if you're really lucky, novelty toys. I presume this all started out as a way to hide the fact that customers are paying a heap of hard-earned cash for what amounts to a greasy sandwich and 27 limp fries. The result has been that for miles around every fast food joint the bins overflow. Such excessive packaging might be explained away on the basis of health and safety – our burger and fries need to be wrapped in three layers of polystyrene to keep the bugs in – but it's product marketing rather than public safety that calls the super-size shots.

Almost a third of all US waste is packaging, around 70 million tonnes of the stuff at the last count, 20 million of which are plastics. In the UK, we produce over 9 million tonnes of waste packaging every year, with over half of all goods, from bread to bananas, being wrapped in plastic. Plastic doesn't grow on trees, though it may often festoon them on windy days. No, this plastic, the millions of tonnes of it keeping our bananas safe from air and lacerating office workers around the world, comes from that oh-so-finite fuel of the modern world: oil. About 4% of the world's oil is used as the raw material for plastics, with another 3–4% going into plastics manufacture. This is equivalent to the total annual oil production of Kuwait and Iraq.

If, like me and the Carbones, you do try to recycle your trash you've no doubt found plastics to be rather a tester. There may be kerbside collections for papers, bottles and cans, but plastics largely remain the ugly bath-time duckling in council schemes. Down at the supermarket there's a row of brightly coloured domes for your glass of every hue, and two huge metal skips with finger-snatching letter boxes for your newspapers and textiles, but plastics? You'll be looking for a long time in some countries. The recycling rate for plastics in the UK and USA is currently less than 5%, and that number doesn't look like shifting any time soon.

It comes back to just how cheap plastics are to produce, and to how varied and bulky they tend to be. While it's often cost-effective for councils and companies to recycle glass, metal and paper, plastic recycling just doesn't hold the same cash incentive. Sorting is a big problem: glass can be separated by colour, so it isn't too taxing. For cans, it's aluminium versus steel – a bit more tricky but if the label's still on or you have a handy magnet then no problem. For plastics it all gets very complicated. There are about fifty different types. The American Society of Plastic Industry, ASPI (Christmas parties there must be a hoot), has helpfully put these into seven groups. For one of the bigger categories, 'High density polyurethane' (shampoo bottles, for instance), every tonne recycled means a greenhouse gas saving of about 1.4 tonnes. But, with caps which can't be recycled on bottles that can, it's not surprising that most plastic ends up in the trash.

On the positive side, plastic collection and recycling is on the up. Today you can go online and order anything from a fleece made from recycled plastic bottles to a bag made out of their caps. Plastics can often fit the 'Reuse' category, in particular all those shopping bags. In the UK a staggering 8 billion bags are given out free each year – 130 for every man, woman and child. The kitchen drawers of the western world are home to many billions of the things, with yet more ending up swirling down streets and blocking drains. This isn't just unsightly: it can cost money and lives. In India, plastic bags have been banned in many areas because of their tendency to block sewers and so endanger health. Many animal species are also at risk. In the waters around Australia errant plastic bags kill turtles, whales, seals and birds who mistake them for food. In Ireland the imposition of a plastic bag tax has resulted in a 90% reduction, the few pence charge meaning most bags are now reused rather than binned.

Reuse of plastic, other than as carrier bags, has unfortunately made a rather stuttering start. The Body Shop, for example, offered refill services so that people could use the same bottles again and again for their beauty products. But only 1% of customers actually used this service and it was withdrawn in 2003.

And now to the things that really shouldn't make an appearance in anyone's bin: the glass and metals. An average Western family gets through about 500 glass jars and bottles each year. The UK as a whole uses around 2 million tonnes of glass packaging each year and recycles about a quarter. This is pretty poor compared to the rest of Europe where the average is over half, Switzerland managing a smashing 95%. Making glass uses lots of heat, and so lots of energy. Recycling cuts back greatly on this manufacturing energy, as well as saving on the extraction of all the raw materials. Every tonne of glass that gets recycled means 1.2 tonnes less raw materials and a reduction of 300 kg in greenhouse emissions.

For metals it's a similar story. Recycling steel packaging, usually in the form of cans, again removes the need for all those raw materials. A tonne and a half of iron-ore and half a tonne of coal is needed for every tonne of new steel, so recycling saves about 70% of the energy it would take to make the steel from scratch. Recycling aluminium holds even more climate benefits: every recycled kilogram saves over 14 times its weight in greenhouse gas from being emitted into the atmosphere – a cut in the energy use and emissions of over 90%.

Another 4% of household garbage is textiles. All those holey shorts, odd socks and itchy jumpers ("Honestly, Mum, I lost it when we moved house") add up to between a half and one million tonnes a year just in the UK. Recycling clothes avoids the manufacture, transport and use of new materials and dyes. And for natural fibres like cotton and wool it also robs the methane producers of yet another tasty, if tickly, snack.

Last of all, in the 'other' category, there are things like the computer keyboard that's had one too many cups of coffee spilled over it; the bits of wood left over from a failed attempt to put up a new shelf; and the commemorative 'Charles and Diana' mug that broke when said shelf fell off the wall. Some of this (the keyboard for example) can be recycled by specialists. Other bits, like the smashed Charles and Di mug, could provide a useful drainage layer at the bottom of a potted plant or might after all find a final resting place as the royal retreat of several thousand landfill bugs. There will always be some stuff that really is rubbish, but well over half of what most of us junk could have a different fate.

Our bin is noticeably lighter now. Through a combination of reduce (no single bananas in moulded plastic for us), reuse (the bursting carrier bag drawer in the kitchen now shuts), and recycle (1,000 composting worms are in the post), we've halved the greenhouse emissions of our trash from almost two tonnes to just one. Such action, multiplied across neighbourhoods, states and countries, can mean big emissions reductions.

Some governments have recognised this and are now taking a very proactive line on household recycling, providing free recycling bins, kerbside collection for a host of materials, local processing facilities and even free composters. In Germany they have taken things even further, imposing fines on those who don't recycle and obliging manufacturers to take back up to 70% of the packaging they produce. The potential is great. In the USA, increasing the current 30% level of trash recycling by just 5% would mean a cut in greenhouse emissions of around 36 million tonnes – more than 100 kg for every man, woman and child.

With the compost bin rather than the trash can overflowing, it's time to grow some food. Planting the kind of fruit and vegetables that you buy regularly and which would often be imported by air – soft fruit and salad for example – is a great way to get your backyard punching its weight in the fight against global warming.

Kate Carbone loves her garden. She cherishes its shifts in colour and texture with the onset of each new season. Her weekends and evenings are largely defined by weeding, digging, sowing and harvesting. When the Carbones first moved into the house the garden was a mess. The borders had merged with the lawns, the thistles and brambles were head-high and somewhere hidden at the back was a leaf-filled pond.

For the first year John and Kate left the garden to its own devices; there were a hundred and one other jobs to do inside before giving a thought to the outside. By the second spring, Kate had redecorated almost every room in the house and was heartily sick of wallpaper and soft furnishings. Having just finished the third coat of magnolia needed to hide what could only be described as the diarrhoea-yellow bathroom, she sat down on the back step for a coffee.

In front of her lay the jumble of brambles and grass, but amongst it, here and there, small flowers were emerging. A patch of what she had assumed was as wild and uncared for as all the rest, was sprouting into life. She plucked one of the new green shoots and breathed in fresh oregano. A quick search among the weeds and she had found sage, tarragon and thyme, plus some very happy looking mint sprouting from an old porcelain sink – Kate was hooked.

That year she, and John, when he could be persuaded, worked their way through the garden, weeding here, turning the soil there, all the time finding old gems to be kept and forming new plans of how to make the garden useable again. The big surprise came late that July when one of the apple trees at the bottom of the garden produced a great crop of 'Lyman's Large Summer', which proved not just edible, but delicious.

On paper the plans were straightforward: beat the borders into place, bring to life the small lawn, dredge the pond, cut back the brambles and open up the herb and adjacent vegetable patch. By the time the nights started drawing in and the frogs in the newly cleared pond had stopped singing, the Carbones still had a mountain of work to do, but they had already made big inroads and Kate now had a whole winter to plan next year's attack. The following year the first seeds went into the vegetable patch and later the Carbones tasted their first harvest. OK – it consisted of three oddly shaped carrots and some marrows so small that calling them zucchini would have been flattering, but they tasted so out of this world that the couple vowed to make next year's harvest bigger and better still.

So it was that order came to the borders and beds. The Saturday morning visit to the local garden centre became a feature of every weekend. When George and Henry came along it was also a great way to get a few minutes' quiet while the kids rammed cake into their mouths in the coffee shop. The big pile of weeds that built up at the bottom of the garden was tidied into a large compost bin which then produced nutritious organic matter to fork into the beds. John, though not quite as enamoured with the plant side of things, took to the role of chief composter with real energy. Each week he could be found checking it to see how much compost he was getting, feeding it with a mix of grass clippings, plant cuttings and the occasional shredded cardboard box, and wagging his finger at the kids when they chucked on indestructible weeds like bindweed or couch grass. One weekend at the garden centre he spotted a wormery and, a sucker for a new bit of kit, was unpacking worms-and-all at the kitchen table an hour later, much to Kate's dismay.

Each year the area of the vegetable patch increased in size, with the Carbones trying all sorts of different things. One year they (and their friends, neighbours and work colleagues) were rather overwhelmed by the pounds of runner beans the garden produced. Another year George and Henry had all but gone on hunger strike after a bigger than expected crop had meant carrot soup/curry/salad/stir fry for lunch for two whole weeks. As the Carbones' experience grew, and the amounts of useful food they produced leapt, Kate began to realise just how much this gardening could help the family out. Money, as always, was tight; hardly having to buy vegetables from the supermarket each week really helped the grocery bill, even if she did blow most of those savings at the weekend on her trips to the garden centre. Now, with the bump starting to kick each time she kneels to pull out another weed, John, Kate, George and Henry get through most of the summer and much of the autumn without having to buy vegetables at all. Though the Carbones today are better off, the thought of home-made apple purée for the new bundle of Carbone joy, rather than the mass-produced supermarket stuff, is a big added bonus. Indeed, Kate has plans afoot to make the garden work even harder.

This year she's going to turn over a section of herbaceous border to blackcurrant bushes and a pear tree. Her ultimate aim is to sail straight through both the fruit and vegetable aisles in the supermarket every week, something George and Henry have always wanted, but to do this without also condemning the family to scurvy. Meanwhile John plans to build a glasshouse. He already has dreams of growing the sweetest ever satsumas and chillies so hot that even Ted from next door will have to cry when he tastes John's chilli dip at the street barbecue. A glasshouse will also help the supply of cucumbers and tomatoes and give them salad leaves right into the depths of autumn, something they miss when dropping temperatures and day lengths mean they can no longer just nip into the back garden and fill a bowl for the dinner table.

Quite apart from the obvious benefits of the Carbones' garden – the exercise, enjoyment and assurance about the quality of the food they're eating – Kate's hobby-come-obsession makes big inroads into the family's greenhouse emissions. As we saw, our food often carries a big climate tag, especially if transported long distances. With an expanded vegetable patch, fruit trees and a glasshouse, the emissions savings will total almost a tonne.

Growing some food in the garden, as well as composting yard trimmings and kitchen scraps, makes a real dent in the family's contribution to global warming. To reduce the climate change-related dangers of spending time outdoors even further, there are several other possibilities. In some gardens emissions can be reduced by planting trees. Aside from the carbon dioxide that the trees suck out of the atmosphere as they grow (every cubic metre of wood locks up around 800 kg of carbon dioxide) they can shade the house in the summer and so help keep the air-conditioning quiet. More garden shade, along with more careful irrigation, less lawn and more drought-tolerant plants, can also minimise water use. Even here in rain-soaked Britain there are plans afoot to build an energy-hungry desalination plant to meet soaring demands for fresh water.

Finally in the garden, there's yet more stuff. As gardens have become 'extra rooms' they've also become prone to the same array of furniture, accessories, gadgets and toys as our homes. Buying less garden tat saves emissions in just the same way as does eschewing a pasta maker for your kitchen or nose hair trimmer for the bathroom. Less stuff equals less embodied energy. Even the choice of the more necessary garden gadgets matters: for those with gardens on the smaller side, changing the lawnmower from a powered to a 'push' model can reduce emissions by another 40 kg a year. Or, like car share, there's always the option of mower-sharing with a friendly neighbour. As for those patio heaters? For every artificially heated outdoor soirée these can belch over 10 kg of greenhouse gas into the sky. That's a third of a tonne for a year's worth of barbecues. Instead, put on another jumper before you go outside. If you need one just write to me: I have some lovely woolly numbers thanks to Nana Reay.

The incentives to curb the greenhouse emissions from our backyards, be it through more recycling and a lighter bin, or by planting herbs at the back door, are big. You may well do most of these things already; if not, maybe the threat of those malaria-swapping, carcinoma-inducing barbecues will spur you to action. If not, here's something that will spur even the most gadget-loving, gas-guzzling sceptic into action: money.

Table 4 Potential for cuts in backyard greenhouse emissions (% of 2 tonnes per household per year).

	Reduce	Reuse	Recycle	Compost	Home-grown
Greenhouse gas savings	Up to 70%	Up to 30%	Up to 30%	Up to 50%	Up to 100%

6

money matters

Get past the first page of any energy-saving booklet and, alongside the smug-looking faces, you'll see the financial benefits of energy efficiency writ large. Long ago, pamphlet writers and policy makers realised that our pockets are a great way to prod us into action. This carrot strategy – insulate your home and you'll save enough money in five years to buy a new car or a holiday in the Maldives – is rife. But, as you've probably spotted, there's a flaw. Less heat wastage in the home equals more money equals bigger car/more holidays/a new fetlock volumiser. It's effectively pollution swapping: exchanging the emissions of home heating for those of longer flights or more embodied energy in the form of stuff.

For every coin we earn, for every free moment in our lives, something has been produced on which to spend that cash and fill that moment. We work hard to be able to while away the weekend in the shopping mall buying stuff. But no matter how hard we work, however big our pay rises or credit extensions, there's always more to buy. The green movement is continually faced with this problem. They tell us to pare down our lifestyles, to reject the consumerism that defines our waking days and save the planet for future generations. But once we've got our composting toilet, introduced the perplexed Merino sheep to the back lawn and installed the hand loom in the spare bedroom, we want to know what else we can buy. Where can we get our retail therapy now?

The answer is for the government-sponsored booklet in the dentist's waiting room to make the wider connection. If we knew that the climate benefits of better home insulation would be wiped out by our buying a bigger car, then we'd more likely spend the money saved on something else. Sustainability is the watchword here and it is down this rocky road that money and climate change are pushing and pulling us with increasing urgency.

Say you improve your insulation. Within five years, you'll have saved back the cost. Or, looked at another way, you'll have saved enough to upgrade your train pass to first class, or to convert your car to dual fuel, or to spend more on local, seasonal food with fewer food miles. Save enough and you could make that down payment on the solar water heating, or even work fewer hours each month. Or, more prosaically, your 'climate dividend' could simply dent your overdraft.

The relationship between money and the climate doesn't stop at making some savings on electricity bills from using low-energy lights. The many and varied ways in which climate change will hit our lives will, in most cases, hit our pockets too. We face a future of paying extra to repair damage to property, cars, gardens and health. Plus there could be soaring taxes, aimed at reducing our emissions and helping governments cope with rising sea levels, failing crops, more severe storms and the rest.

Over in Alabama at Casa Carbone there are big changes afoot, not least the arrival of someone whose generation will be lumbered with the global heating bill we're busy running up. Kate Carbone, as so often, is in the garden. Her full-term cotton-covered bump is obscuring the weeds at her feet and her back is aching like hell. Slumping down on the back step Kate admires her thriving plot – the rows of carrots and swelling pumpkins that, mashed to an orangey pulp, will make perfect baby fodder. Such sunlit dreams of feeding time and carroty faces will have to wait though – the back ache's worse and these contractions have never heard of Braxton Hicks. Time to phone John and get panting.

Twelve hours and 16 high-stepping laps of the maternity ward later the Carbones have a very cross, but healthy, daughter. Grandma Carbone brings in George and Henry carrying, respectively, a bear four times the size of the baby and an activity centre with 'real train noises'. As the Carbones gather around the pale, smiling Kate and the well-wrapped baby, the question of the moment is "What's she called?". Suggestions range from Ruby (Grandma Carbone – "it was your Great Aunt's name"), through Belinda (George – "My rabbit's a girl too"), to Buffy (Henry – "I'm 8 and she's my hero"). John and Kate are already decided though: Lucy Carbone has arrived.

Lucy has been born into an affluent family in the richest and most powerful country on the planet. She has opportunities that babies in the developing world can only dream of. Nonetheless, she too faces a lifelong battle with intensifying and accelerating climate change and a powerful sense of disgust at the knowing profligacy of her elders. If kids ever needed an excuse to hate their parents then Lucy Carbone and her contemporaries have a cracker in global warming.

She doesn't know it yet, her thoughts currently being centred more on nipples than work, but Lucy is destined for a glittering career in banking. Cut to 28 years hence (apply wobbly time travel effect as appropriate) and we find our Lucy at work as a team leader in the PR section of a thriving multinational bank. At 28 years old Lucy is still surrounded by those who did the job when every week brought another trans-America trip to meet the San Francisco team or red-eye flight to liaise with the Zurich branch. Now, in response to soaring air travel costs and the ease of virtual conferencing, Lucy rarely steps on a plane, with weekly international meetings occurring from the comfort of her own desk.

Outside Lucy's office window the parking lot has changed almost beyond recognition. The cars are smaller than today's and there are fewer of them. The area for car parking has been pushed back towards the boundary fence; the space close to the building has been turned over to cycle racks. Bike-buying grants started rolling out over a decade ago and for those living too far away to cycle, the car-sharing scheme is a must – arrive alone in a car and the already steep parking charges are doubled.

Even the building Lucy works in is different. It looks just as shiny and utilitarian as the corporate office blocks of today, but beneath the surface its construction is in line with much higher energy efficiency standards. Walls and floors have top spec. insulating properties, windows are situated to maximise passive heating and cooling, and even the revolving front door is designed to keep the temperature stable. No crumpled recycling trays in the bins here. Now each office in Lucy's building must by law have a recycling and energy-saving policy. Every floor has an energy officer and faces a green audit as part of annual accounting. Thermostats and lighting levels are checked and optimised, electricity and paper-saving options on office equipment are set as defaults, and anyone repeatedly throwing recyclable stuff in the waste bin gets a caution.

As the days are crossed off a dozen calendars around the office and the talk around the low-energy water cooler becomes dominated by holiday plans, yet more transformations are apparent. Jet-setting holidays are now off the agenda for most people, as a combination of emissions taxes and increased concern about global warming boosts tourism at home. It's not only climate-awareness that has changed the global tourist map. Holidays, even before Lucy was born, were changing for the Carbones. By the time it's her turn to flick through the online brochures, some of the destinations we currently flock to have dropped off the tourist radar altogether, the remnants of their heyday now little more than fading postcards.

Just as Mexico – the destination of choice for millions of sun-seeking Americans – got too hot for the Carbones, so the European equivalents faced temperatures that forced the tourist lifeblood of the economy ever further north. Jet travel, initially seen as a boon for tourism, has for many become a curse. The very same jets that in the 20th century flew in millions of visitors were at the same time driving the climate change that now keeps the tourists away.

Coral reefs once lured millions to resorts in the Caribbean, Australia and south-east Asia each year, along with billions of tourist dollars. By the time Lucy fancies a holiday underwater, many of these reefs and their resorts have disappeared owing to rising sea temperatures, over-fishing and pollution.

Today, dive tourism brings $2 billion a year to the Caribbean. Add to this the value of the reefs to the fishing industry and as protection from the sea, and these crusty wildlife havens earn the area between $3 billion and $4.5 billion every year. By 2015 it's estimated that these revenues could be tumbling by almost a third of a billion dollars each year, as tourists bypass the Caribbean for less degraded beaches on which to strut their stuff.

Down under, the Great Barrier Reef currently earns the Australian tourist industry over AUS$4 billion. Here the same threats to the coral, its wildlife, and the livelihoods that depend on it are apparent. Some scientists predict that up to 95% of all living coral in the reef will have disappeared by the middle of this century. The economic costs of such a die-back have been estimated at AUS$8 billion dollars and 12,000 jobs by the year 2020.

In his book *Last Chance to See*, the late great Douglas Adams visited numerous animal species on the brink of extinction and highlighted the need for their protection. Such action is vital for ecosystems and endangered species, and for the people who depend upon them. Just as reefs bring tourists and their wallets, so too do the iguanas of the Galapagos, the kiwis of New Zealand and the orang-utans of Borneo. Concern over damage to many ecosystems and the disappearance of animal and plant species has spawned the whole new holidaying ethos of 'ecotourism'. Faster than you can say 'handmade dominos', ecotourism has grown into an industry worth billions of dollars a year. The basic premise is that those visiting rainforests, savannah and the like should ensure the protection of the ecosystem and of the local economy. You therefore return with photos and stories, rather than the more traditional lump of dyed coral, mass-produced orang-utan doll, or slowly decomposing elephant's foot stool.

Unfortunately, rapid rises in sea level, temperature and rainfall won't be stopped by all the locally produced didgeridoos or low-impact nature walks in the world. It has been estimated that climate change could push up to a third of all plant and animal species to extinction during the 21st century, from the gibbons whooping through the forest to the wriggling bugs they eat. Whole ecosystems will be undermined with unknown impacts up and down food chains, of which we are usually part. The millions of people who rely on the flow of tourist money brought in by local wildlife will see their livelihoods disappear along with the last lonely cry of that final gibbon.

No Borneo jungle or St Lucia diving holiday for Lucy then. How about a ski trip? If she goes high enough, or the resort has enough snow machines, no problem, but many resorts that people visited in droves when Lucy was born are long gone by the time she collects her first lift pass. In some places that once saw a reliable 40 inches a year, snow has quite literally become thin on the ground.

Each winter I spend a lot of time flicking from one weather forecast to another, trying to find one that predicts snow for West Lothian. Many are the nights I can be found peeking out between the curtains to see if the 5% chance of snow predicted has given me an excuse for a day at home, drinking hot chocolate and building snow forts. Invariably it hasn't. This may annoy me and my wife, who has to put up either with a cold-footed husband moaning about climate change or a half-demented husband dancing around in celebration of a few snowflakes. For many in Scotland though, the failure of winter to bring any reliable snow spells unemployment. The ski industry in Scotland was never massive, being worth around £30 million a year at its peak. But what there is is now shrinking at an alarming rate. During winter 2004 – another characterised by little snow and by midge bites in January – two of Scotland's five ski resorts finally gave up the battle against falling revenues and rising temperatures and put themselves up for sale. The situation for the other three resorts looks bleak too and the industry as a whole is unlikely to last another 20 years. For the hundreds of workers in the Scottish ski industry, job losses, like snow-free winters, seem inevitable.

Worldwide the picture for the ski industry is equally grim. In Austria the snow line may rise by more than 300 metres in the next 30 years. On the other side of the world in Australia it's likely that by 2070 none of the existing ski resorts will be viable. Between them they net around AUS$400 million a year. In Switzerland, home of the chocolate box chalet and après-ski bill large enough to bring on snow-blindness, over 60% of the 230 ski resorts could find their snowfall becoming unreliable. Over in Germany and Italy it's a similar story.

Of course, as long as we all still want to go on holiday, someone somewhere will reap our tourist dollars. If the reefs are allowed to degrade too badly in one area but are better protected in another then we'll switch. In 1997 the Alps had little or no snow on the slopes. Instead, Morocco saw a huge surge in visitors to its ski resorts. Scotland's ski industry is going rapidly downhill, but warmer summers coupled with the Mediterranean becoming uncomfortably hot may simply shift the tourist boom to the middle of the year.

The tourist industry is not the only one facing great uncertainty. Severe weather can ruin farmers within a matter of minutes. Ice storms, like those in the USA and Canada in 1998, may kill many hundreds of cattle. Drought, like that resulting from the 1997–98 El Niño, can wipe out entire harvests – in New Zealand alone it cost well over half a billion NZ dollars.

If working the sea, rather than the land, calls, then a career in sailing might well be a better option than fishing. There's no denying that systematic over-exploitation has caused many of the crashes in world fish stocks and attendant job losses. But in many instances, North Sea cod for instance, rising sea temperatures make such crashes more rapid, and potentially more permanent. Separating the effects of over-fishing and global warming is difficult. But together they have led cod stocks to fall from over 250,000 tonnes in the 1970s to less than one fifth of that amount today. In the UK, 15,000 fishermen face the prospect of cod becoming commercially extinct in the next five years.

Foresters too face an uncertain future. More carbon dioxide should help some trees grow more quickly, but will also bring more pests, sustained droughts and the potential loss of large areas of forest due to storm damage. An average temperature hike of 2 °C during this century would push the ideal range of many US forest species about 320 km north. Such shifts are fine when species have millennia to colonise the warming north and retreat from the baking south. Given less than 100 years they'd need to be migrating north at over 3 km a year – way beyond what many trees can manage.

Tree surgeons meanwhile will be turning work away. In its 'great storm' of 1987 the UK lost an estimated 15 million trees. Kent's famous village of Sevenoaks (one-time home of 'Bloomsbury Set' writer and garden designer Vita Sackville West) became 'Oneoak and some rather large logs', and the nation's tree surgeons got a year's worth of business overnight. Builders should also be in the money. There will be a lot more repair and construction work, from subsidence damage and buckling roads to new flood defences and energy efficiency improvements.

John Carbone's line – insurance – could be an extremely rocky one depending on the people and assets a company insures. At the moment, the average person in the USA spends over $2,500 each year on insurance premiums. Between the mid-1980s and the late 1990s, weather damage, from storms and floods to tornados and drought, cost an estimated $253 billion, with 40% of this being privately insured. These already huge costs are rising fast. Climate change is playing a part, but so is wealth. Each year there are more people with more stuff in the way of more severe weather. If insurers are essentially gamblers, then the odds are getting worse and worse. Given the high standard of living in the USA, and the vulnerability of many US citizens, the potential financial losses are huge. Add together the value of properties along the Gulf and Atlantic coasts, from the beach condos to the Taj Mahal wannabes, and you've got well over $3 trillion of real estate sitting next to a swelling sea.

In 1990 flooding hit 23,000 people along the coastlines of Europe. If we fail to take action, then by the 2080s we will see around 5 million people a year suffering the same fate. A 1 metre rise in sea level would flood the homes of over three and a half million people in the Netherlands – a quarter of the population. It would cost $186 billion and destroy over 200 square kilometres of land.

The insurance offices of the world are having an increasingly bad time as weather-related claims soar. Between 1969, when John Carbone started work with his current employers, and 1998, 650 US insurers went bankrupt, 50 of these as a direct result of natural disasters. Hurricane Andrew in 1992 and the ice storm of 1998 cost them dear and spurred efforts to better model disasters and plan the siting of buildings, abandon high-risk insurance, or hang up those pinstripe suits altogether.

The costs of the 1998 ice storm were huge. In the USA and Canada it led to over 800,000 insurance claims to a value of more than a billion dollars. There were 45 deaths, over 5 million people were left without power, and 17.5 million acres of US forest was damaged. Workers in Canada lost a billion dollars in wages, with the agricultural sector haemorrhaging a further $25 million in livestock and crop losses.

In Europe, the story is similar. Severe storms in the 1990s clocked up heavy insurance losses. 'Daria', in 1990, cost France, Germany and the UK almost $6 billion and killed 95 people; the very next month storm 'Vivian' crashed its way along the North Sea coast, costing $4 billion and killing 64. Floods too have taken a heavy toll: the 1996 Odra flood in central Europe killed over 100 people and caused $5 billion worth of damage. Global economic losses from catastrophic events rose ten-fold from the 1950s to the 1990s. A single heavy hailstorm that struck Sydney in 1999 cost around a billion dollars.

These severe events that keep insurers awake at night are set to become more frequent and more intense. In response, premiums have already increased, with some areas becoming uninsurable because of repeated flooding. Even householders in low-risk areas might soon face having to pay excess to get cover for extreme weather.

The terrible thing about rapid climate change for insurers is that it doesn't fit with how they've usually done business. It used to be enough to ask "How often do we have storms with hailstones the size of baseballs?" and set the premium for car windscreens accordingly. When formerly 1-in-50-year or 1-in-100-year events may start to occur every year, or in areas where they've never been seen before, it all gets rather trying for insurers clutching their historical event tables.

Governments are also footing big bills and so, through taxes, are we. The Canadian administration spent around $15 billion on disaster relief between 1982 and 1999, while in the USA such payments amounted to $119 billion between 1977 and 1993. Hurricane Andrew cost over $15 billion as it ripped through the south-eastern USA in 1992. The same hurricane today would cost double, with inflation adding to the fact that a lot more property and people would be in its path.

By the time Lucy is looking for her first home, her choice is led much more strongly by climate change than the first-time buyers of today. "Is it in a flood risk area?" is something she asks even before looking into the number of bedrooms. Her next questions are: How prone is the area to storms? How hot does it get in the summer? Is the area home to West Nile Virus/ Malaria/Chagas disease? For a property free of all these Lucy will pay handsomely.

Today, even homes in the safest locations face costly increases in buildings insurance. In the UK the creeping cracks of subsidence already cost over $600 million a year. The Association of British Insurers predicts that the figure will rise to $4 billion a year by the middle of this century. A real deal-breaker question for home-buyers of the future will be "How is it heated and lit?". With electricity, oil and gas prices at all-time highs, no first-time buyer will afford a house with poor insulation or inefficient heaters and lights. When Lucy goes looking, solar panels aren't some estate agent oddity, but a standard fixture akin to an *en suite* bathroom.

Several governments have already introduced grant schemes designed to encourage more climate-aware living. These include help with the upgrading of heating systems, insulation, solar power, dual-fuel car conversion and buying compost bins. In the future we are likely to see more subsidies for renewable energy use, low-waste households, public transport and bicycle use, and more energy-efficient white goods – some carrots to go with the many sticks.

The aisles of the local supermarket in 2030 will likely look very different. Hand-tied chives? Take a hike. Hopefully, the food miles of each item will be labelled. Increased taxes on airfreight will mean if you want fruit and vegetables out of season from the other side of the world you'll have to make do with ocean-shipped produce, and pay a premium for it. The plastic-sheathed bananas of today should have gone, along with the triple wrapping on meat and dairy produce.

At the checkout there's no rack of carrier bags. If Lucy doesn't bring her own she pays through the nose for decomposable bags (refund on return). And her grocery bill? The estimates for the impact of global warming on worldwide food production are, given the uncertainty involved, wide ranging. The broad consensus is that if we see a global temperature rise of 2.5 °C or more then, as well as a multitude of other problems, we can all expect the cost of food to rise, with cereal prices surging by up to 45%.

Back home, having eaten her food-mile-lite dinner, there's less trash on the worktops and no throwing what there is into the wheely bin. Lucy separates plastics, metals and paper for recycling and sorts her leftover food for composting or collection. Too much in the bin at the end of the week and a fine is taped to it the next day. In fact, this already happens in some parts of the USA and Europe. 'Pay As You Throw' households shell out for every bag of waste. These schemes, though in their infancy, typically increase recycling rates by 30 to 60%.

As climate change intensifies, the financial carrots and sticks it creates will alter our lifestyles and pound our pockets. What though, if we could bill people for their emissions? What if we could charge the owner of the SUV next door, or the jet-setting advertising executive down the road, for the extra burden they place on our climate?

Imagine, for a moment, their hallways. The post has been delivered and there, amongst the usual heap of junk mail, lies what is obviously a bill. This is no ordinary bill: it is a demand for the damage they've done, it's an invoice from countless people, most of whom they've never heard of, in countries they've never visited: this is their own personal emissions statement.

The widely accepted range (accepted because it's so wide) is that every tonne of greenhouse gas emitted causes damage costing between $2 and $80. This is not terribly precise. For argument's sake, let's take the midpoint of these extremes, and say that every tonne of greenhouse gas costs $40 through flood and storm damage, crop failures, increased health care and the rest. The Carbones are responsible for over 39 tonnes of greenhouse emissions annually. This adds up to around $1,500 of damage every year for this typical US family. Simply driving an SUV takes these so-called externalities (nothing to do with bumper stickers or hub-caps) up an extra $500 a year.

What then might the world's big corporations be liable for? Recent studies have made a direct connection between human greenhouse emissions and weather-related disasters, such as the 2003 heatwave in Europe. If, for instance, the more than 20,000 deaths that resulted from this scorching summer were linked to the global warming contribution of

Table 5 The annual Carbone emissions bill (in $ worth of planet damage).

	Cars	Air travel	Home energy	Food
Carbone emissions bill				
$1,500	$720	$100	$520	$180

some of the world's largest greenhouse gas emitters, litigation lawyers everywhere would be totting up one hell of claim.

Using the estimate of $40 per tonne of greenhouse gas, the annual costs arising from the emissions of some corporations run into many millions, and in some cases billions, of dollars.

Who pays this bill at the moment? In most cases it's who-ever's house is flooded, their insurer, or their frantic govern-ment and sea defence building programme. Some of these costs will come back to bite the SUV driver and the jet-setting executive through increased taxes. But the global nature of cli-mate change means that the bill is likely to end up on the door-mat of someone much poorer.

If the Carbones' emissions cause $1,500 worth of damage, should they be taxed accordingly? In recent years the idea of giving individuals emissions quotas has emerged. Though still mocked as impractical, ineffective or even communist, it is beginning to catch on. There are various versions of the idea in circulation, but essentially they boil down to the same thing: each of us – you, me, Kate and John Carbone, even Lucy – is given a climate-warming allowance, and every time we fill the petrol tank of our car, hike up the heating, or go on a far-flung holiday we use up some or all of that quota. The idea is that by setting such rations, taxing those who over-emit, and reward-ing those who don't, our collective emissions will be driven down and some kind of global equity can be reached. Climate credit cards have been suggested, the price of each tank of petrol being debited from your bank and your climate account simultaneously. Need to go over quota? That's going to cost you – say $40 for every extra tonne of greenhouse gas. Keep under quota and you get to sell your unused climate credit and spend the money on a new hair shirt...

The idea's great in theory and I'd love to see it become reality. Getting developed-world governments to commit to such direct taxation of an individual's emissions seems a long way off though. If a handful of disgruntled lorry drivers can bring the UK to a standstill over high fuel duty, as they did in the autumn of 2000, then imagine the reaction in the USA to the news that, instead of 'No New Taxes', everyone will be getting carbon dioxide quotas. Under the ground-breaking Contraction and Convergence model, designed to reduce emissions by the magic 60% by the middle of this century, we would each be allowed to emit 400 kg of carbon a year. Just one long-haul flight would blow a year's allowance. Carbon quotas for individuals will have their day, but if they are to be effective, the truck drivers and everyone else will need to know exactly why they're being made to pay.

From a political perspective new taxes, and increases in existing ones, are inevitable as the effects of global warming begin to bite. Such carbon or climate taxes, designed to discourage energy wastage and fossil fuel use, are likely to be extended from business to include communities and individuals. From extra payments on air travel to charges for more than one sack of rubbish each week, the average family will be under more and more financial pressure to cut greenhouse gas-producing activities. But whereas individual climate-saving actions generally cost nothing or actually save us money, some national or international mitigation measures will be very expensive. The real poser for governments is whether the costs of the destruction that climate change will wreak tomorrow will outweigh the costs of cutting emissions today. For many countries it's obvious: Yes, cut them and fast.

Sea level rise could effectively bring the Netherlands, for example, to its knees. Without mitigating climate change the only other option, besides evolving gills, is for the Dutch to build massive sea defences. The cost of adaptation to such sea level rise is colossal, at over $12 billion. In England and Wales there are currently more than one and a half million properties at risk from flooding. If flooding goes unchecked, instead of the $2 billion currently lost each year, we'd be looking at 20 times this figure by the year 2080.

In Japan, a rich country characterised by cramped low-lying and coastal settlements, sea level rise has similarly devastating potential. A hike of a metre – something we may well face as the 21st century draws to a close – would put at risk over 2,000 square kilometres of this already space-starved nation, endangering over 4 million people and 109 trillion yen's worth of property. The list goes on, the numbers huge. For example, Egypt would face a $35 billion bill from lost tourism, land and property with a 50 cm rise, while Poland would lose $30 billion from a 1 m sea level rise. For Australia and New Zealand the current prediction is for a drop in GDP of between 1% and 4% with a doubling of carbon dioxide. Globally, a doubling of carbon dioxide in our atmosphere may cost over $300 billion a year by the middle of this century.

The Kyoto Protocol is the scarred battleground, where economics and science have been used to suit every interest. This protocol represents the first real step in joined-up international action on climate change. For rich countries with large emissions, the USA and Australia for instance, even the rather limited reductions set out in the protocol would mean some big spending.

Back in 2001 the USA withdrew from Kyoto citing the economic cost it represented as a key reason. The cost of achieving the Kyoto-sized cut that the USA rejected (7% of its emissions in 1990) comes in somewhere between $13 and $397 billion.

So how do the mitigation costs stack up against climate change-driven damage in the USA?

You'd assume that the former must be far in excess of the latter to sustain the US position. Not so. For a doubling of pre-industrial carbon dioxide concentrations, losses of almost $70 billion a year have been estimated for the USA of 2050, health costs alone topping $16 billion a year. If some of the more dire scenarios come to pass (a sea level rise of several metres due to collapse of the West Antarctic ice sheet for example), costs to the USA of many hundreds of billions of dollars are possible. Both action and inaction will therefore present a big bill. A recent study which weighed these two bills against each other concluded the net costs to the USA of complying with its Kyoto commitments would be insignificant (±1% of GDP).

The arcane world of climate change economics will undoubtedly continue to provide ammunition for politicians of every persuasion, especially given the huge ranges of the forecasts. Predictions will get better with time, the ranges will narrow and the financial arguments for action are likely to get ever stronger. Unfortunately, by the time economists have a precise figure on the cost of climate change the flood water will probably be lapping at politicians' doors. While they sit on their hands and wait for the worst, we can get on with tackling global warming in every area of our lives, without requiring international consensus or kow-towing to vested interests.

Can we show our politicians the way? The scientific community warns that a 60% cut in greenhouse emissions is required to avoid catastrophic effects, yet even with a properly working Kyoto Protocol, global emissions will fall by just a few per cent. How then do our own actions stack up alongside the 60% target? Over a lifetime, from how we travel to nursery school to the type of funeral we have, just how big is our own contribution to global warming and how big a dent will climate awareness make?

7

a green inheritance

One of the most distressing things about climate change is its momentum. Much of the greenhouse gas we emit today will go on affecting our planet for centuries. Molecules of carbon dioxide from today's shopping trip or journey to work will still be knocking about in the atmosphere when our grandchildren's grandchildren first emerge into the warmer world of the 22nd century. To use a deliberately ironic metaphor, the effect of all the greenhouse gas we are adding to the atmosphere, day in, day out, is like never-ending snowfall at the top of a glacier. The more snow that falls, the bigger the glacier gets and the more far-reaching its impacts. So every kilogram of carbon dioxide, every extra jet flight or office light, is another flurry to join the world-changing glacier of global warming. The big difference, of course, is that most of the world's real glaciers are shrinking precisely because our greenhouse 'glacier' is growing.

Over a lifetime, each of us clocks up a lot of greenhouse gas, but some are responsible for an awful lot more than others. The sum of every car journey and jet flight, every air-freighted kiwi fruit or plastic wrapped banana, is the global warming legacy we bequeath to all those who come after. Cutting these lifetime emissions, this extra snow on the glacier, holds the key to leaving a greener inheritance.

Just as Lucy Carbone is starting life, with all the decisions about her own lifestyle still to make, Grandma Carbone is reaching the end of her days. Born between the two great wars she has seen the rise in car ownership, home energy use, food consumption and technology that have transformed our world. These same shifts have also been largely responsible for the rapid climate change her granddaughter's generation faces. Grandma Carbone can recall a time when cars were still a novelty, jet travel was yet to be invented, and when the only things that went into the trash after a meal were some well-chewed chicken bones if it was Sunday.

We can't simply go back to a pre-1950s way of living; indeed, few would want to. It's all too easy to don the rose-tinted spectacles and hanker after a time when life was simpler. But this would be to ignore the great leaps forward in technology, healthcare, diet and standard of living that we now enjoy across much of the developed world. On the other hand, the price of this breath-taking development has been a huge growth in greenhouse emissions.

Grandma Carbone was born into the pre-war America of 1932. For the first 18 years of her life she went to school, helped out at home and saw numerous men in her home town, including her own broad-shouldered older brother, leave to fight and never return. The war years were hard, emotionally and economically, for everyone. They were, however, marked by relatively low household energy consumption, lots of home-grown and locally produced food, and no family car. Hence her childhood emissions were very low: just a few hundred kilograms per year. On her 18th birthday Lucy's Grandma met the amiable Lt. Francis Carbone at a dance, fell in love, and within six months was married.

In the hot summer of 1950 the newly married Grandma Carbone moved with her husband into their new house. In their first car – a blue and white Lincoln Capri loaded with suit-cases – they pulled into the weed-choked drive of a clapboard house, and so began over forty years of married life – years which, as elsewhere in the world, saw huge changes in travel, home energy use and emissions.

At first Grandma and Grandpa Carbone produced relatively little greenhouse gas. Their car was inefficient, but the average annual distance travelled in the 1950s was substantially less than today. Their Lincoln clocked up less than 6 tonnes of greenhouse gas per year, 3 tonnes for each Carbone. In this first decade of marriage there were no plane journeys; even on their honeymoon they took the train to the Rockies.

The house was draughty but for these 1950s sweethearts the 'more clothes' solution was accepted without a second thought. In Alabama's hot summers they did less work at midday and kept in the shade. Coal-fired heating represented their main source of greenhouse emissions, amounting to about 5 tonnes a year. Household electricity use back then was about a tenth of the average today.

Most of the Carbones' food was locally produced and, thanks to tight budgets and wartime attitudes to waste, their trash was about a third of today's average. Throughout the 1950s then, Grandma Carbone's emissions were low, at about 6 tonnes a year if we split the household equally between her and her husband.

By the end of the 1950s and into the 1960s, things were a-changing. Electricity started replacing coal and wood as the home energy source of choice in the USA. Mains electricity supplies spread out across the country, bringing electric lighting, heating and power for the washing machines, fridges and other white goods that began to appear in the windows of the town stores. With greater supply, low prices and soaring demand, total energy consumption went up and up.

By the mid-1960s energy use in US households had doubled from that when the Carbones first moved in together. Electricity use quadrupled over the same time. Into these heady days of technological progress and nightmarish Cold War threats John Carbone was born, adding his own, initially small, contribution to the greenhouse emissions of the Carbone household.

In the 1970s the energy revolution was in full swing. Household emissions across the developed world soared. By 1975 home energy use in the USA had leapt to three times that in 1950. The Lincoln Capri was traded-in for a more stylish family car with an engine even less efficient than the family's 1950s roadster; it did fewer than 14 miles per gallon.

With each passing year, the greenhouse emissions of Grandma Carbone kept on rising. Labour-saving devices appeared in her kitchen, as in others. The fridges and washing machines of the 1960s were just the first scouts in kitchens, leading the way for today's army of tumble dryers, dishwashers, and eventually bread makers and fondue sets. By the time in the late 1980s that John Carbone left his family home, it, like many US houses, had air-conditioning, a microwave, two refrigerators and three TV sets. All around the house little red standby lights had taken up residence, the tell-tale sign of this appliance-invasion steadily eating its way through kilowatt hour after kilowatt hour of electricity.

By this time Grandma Carbone and her husband were now travelling a lot further by car, jetting to Mexico for holidays and buying more exotic food. Even their once-echoing rubbish bin was filling up fast.

When, in the late 1960s, the young John Carbone first escaped the watchful eye of his mother and slipped out of the front door into the quiet lane outside, seeing a car was still quite a big event. Across the country the number of car owners amounted to 70 million – a select club compared to the 200 million cars on US roads today.

Imagine a time-lapse camera set up outside the Carbone house, trained on that dusty track of the 1950s. Its first shot is the pink-cheeked Grandma Carbone and her husband drawing into the driveway, full of dreams of their married life to come. Now, speed through the years. To begin with, the road sees just the occasional flash of a car; then the surface goes from dirt track to tarmac and the flashes of cars come thicker and faster. By the 1980s, the cars on the road start to blur together – the road has been widened and the traffic roar is increasing. Once it was a lane to ride bikes along, whose puddles served as oceans for paper boats; today, even playing near the 'lane' is a no-no. Now the blur of cars continues into the night.

The same insidious change would be seen had we pointed our time-lapse camera into the sky. A clear blue canvas to begin with – the skies over Alabama home to just a handful of planes each day and, before the war, no jets at all. Then, as the years roll by, more and more streaks start criss-crossing the sky as air travel begins to boom. Speed our camera up and it looks as though a three-year-old has been let loose with a box of chalks. The rate of growth in air travel since the 1950s has been astounding. Today there are 19,000 airports in the USA alone; together they see 9 million departures every year, subjecting over half a billion passengers to looped in-flight movies and vacuum-packed trout à la crème.

The Carbones' main airport, Birmingham, is not exactly JFK. In 1931 it was headline news when the airport was visited by its first passenger plane. By the mid-1980s it had around 40 departures a day. Then came cheaper fares and flying as the norm for both work and holidays. By 2000, Birmingham airport was sending over 80 departures into the skies of Alabama every day. Globally, this kind of air-traffic growth has been repeated again and again: well over a billion passengers now zip about on the world's airlines each year (Figure 14).

Go to the quietest spot you know, somewhere a long way from the nearest roads, where nature still seems to hold complete dominion and the sheer wildness of it all sends a shudder down your spine. Now look up. Chances are you'll find you're not alone. High in the skies above there will be a jet packed with business people, holidaymakers or hand-tied chives, marking its route with a long white vapour trial, and laying out its greenhouse gas nets like it's the last day of the fishing season.

You get a real sense of just how many flights are passing overhead at any given moment from the effect that the terrorist

All flights for 09/03/01 for Hour 0 Above 25,000 ft.

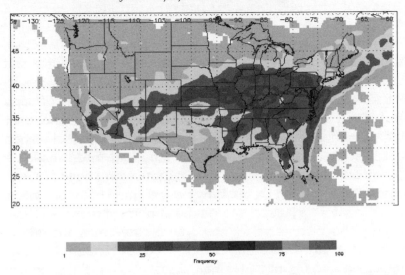

Figure 14 Flight frequency over the USA, 3 September 2001 (Patrick Minnis, NASA Langley Cloud and Radiation Group, USA).

All flights for 09/11/01 for Hour 23 Above 25,000 ft.

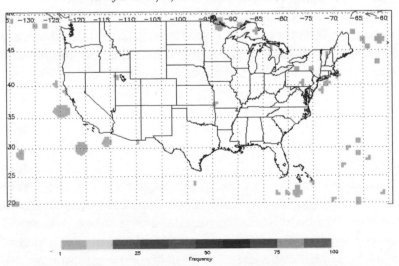

Figure 15 Flight frequency over the USA, 11 September 2001 (Patrick Minnis, NASA Langley Cloud and Radiation Group, USA).

attacks of 11 September 2001 had on a satellite's view (Figure 15). The skies were suddenly emptied of the thousands of jets normally making their way across the USA. For a brief window, while all air traffic was grounded, the skies became clear and sombrely quiet.

Back to the climate impact of Grandma Carbone's life and the 1990s. She and her husband retired in 1992 and their emissions finally started to fall. No more trips to work meant a cut in their annual car travel. But they had more spending power and more time. Why have one jet-setting holiday a year when you can have three? Why have a small runabout when you can afford a Jaguar? More time at home also means more energy use there. Overall, the greenhouse emissions during retirement for Grandma Carbone and Grandpa Carbone remained large. Only after her husband died did emissions really start to drop back when Grandma Carbone swapped the gas-guzzler and draughty house for a nippy runabout and a well-insulated retirement flat.

For Grandma Carbone the picture is almost complete. After more than 70 years of life, most of it very happy and all of it in a fast-changing world, there is one final (and potentially big) addition to her lifetime's climate legacy.

The Carbones are in mourning. Two nights ago the phone rang and a softly spoken casualty nurse told them that Grandma Carbone had been rushed to hospital after a stroke. The ambulance crew had done everything, but she had been declared dead on arrival at the hospital. After the initial devastation, the Carbones are slowly coming to terms with her death and their minds are turning to arrangements. Already, John has contacted a local funeral director to discuss the service and burial, all the time fielding calls from friends and relatives.

Grandma Carbone had never talked about wanting a specific type of interment, though she regularly said that the last thing she wanted was a lot of fuss. In the end the family opts for a traditional send-off. They arrange for the so-called 'Presidential' option. This all-inclusive package comprises three limousines to take close relatives to and from the service, four pall bearers, a top-of-the-range coffin, and a steel-lined concrete vault, with a 50-year no-rust guarantee.

This final steel-clad send-off for Grandma Carbone sends one last large belch of greenhouse gas into the waiting skies, the final kilograms of a lifetime's climate bequest.

Given that death comes to all of us, opting for a climate-aware funeral is one way we can guarantee a greener inheritance. Our burial is the last chance any of us will have to either improve or worsen the climate we leave to those left behind. Funerals may not appear to be that big an issue climate-wise, but once you enter their velvet-curtained world, the huge amounts of material and energy used show just how important this final choice can be.

The Valley of the Kings has nothing on the sheer mass of material we bury each year. In the USA alone, over 1.5 million tonnes of concrete and 14,000 tonnes of steel is sunk into funeral vaults. A further 90,000 tonnes of steel goes into funeral caskets, along with nearly 3,000 tonnes of copper and bronze. Remember embodied energy? Burials in the USA produce over 1.5 million tonnes of greenhouse gas every year – equivalent to the emissions of about 200,000 big cars. All this comes atop the environmental damage resulting from the embalming fluid (800,000 gallons per year) and hardwood (30 million square feet) used in US burials each year.

In a world where success is often judged by the size and expense of the car someone drives, it was inevitable that more and more of us would demand the Rolls Royces of the hereafter. Casket makers haven't let us down. At some point most of us will be faced with a brochure detailing the ways in which we can give our loved ones the send-off they deserve and, in the small print, how this will bankrupt us. Glossy page after glossy page describes boxes with names like 'The States-man' or 'Excelsior' with the blurb stating just how thick the brass plating is, what flowers are embroidered on the silk inte-rior, and whether the things come with an air- and watertight seal as standard.

Let's push the boat out. What can we get that says to every-one else in the cemetery "My family thought so much of me they remortgaged their house just to pay for my funeral"? How about a 'Sultan'? A snip at $6,000, this solid bronze number is lined with crushed velvet, has a seal guaranteed against rust, water or air entry for 50 years, and even comes with a discount on the 'Cambridge' vault. The vault? Well, the 'Cambridge' offers true quality and peace of mind, it being earthquake-proof up to a Richter scale of 5.4. Its steel-reinforced concrete construction, over a tonne in weight, comes complete with name plate and water seals. Inside, it complements the pol-ished bronze of the 'Sultan' beautifully with its own thick bronze lining, just $10,000 all-in.

And the climate tag: over a tonne of greenhouse gas.

If you really want to Rest In Peace there is a lot of weighing up to be done. Both the standard options – cremation and tradi-tional burial – have positive and negative aspects, climate-wise. Cremation using natural gas uses about 20 gallons of gas, while that with fuel oil gets through about 30 gallons per person (the research for this chapter wasn't a barrel of laughs), leading to emissions equivalent to about 120 kg and 300 kg respectively. Added to this are the emissions arising from the burning of the body itself. The amounts are relatively small at around 120 kg of greenhouse gas per person. In total, having your body cremated is likely to result in about half a tonne of greenhouse gas. Not great, but still better than the 'bells and whistles' burial. But cremation has the big drawback of causing air pollution. Mercury and dioxins from crematoria are now strictly controlled in most countries.

Burial, on the other hand, avoids air pollution problems and can be the more climate-friendly option – if the embodied energy involved in elaborate caskets and chambers is sidestepped. 'Green' burials now account for a significant and growing percentage of funerals. These aim to reduce the overall environmental damage, with the impacts on air, water and soil all being considered. For an alternative casket, options include biodegradable body bags, cardboard or bamboo coffins, and wooden caskets sourced from renewable forestry. Such burials turn the premise of hermetically sealed vaults on its head. Instead of entombing the casket inside a thick, concrete-lined vault to prevent any soil-based decomposition for generations, green burials are designed to allow the straightforward breakdown of both casket and body in the soil.

Grandma Carbone's life was one of rapidly increasing emissions as living standards and energy use across the developed world rose following the spartan years of the Second World War. Her lifetime and in-death emissions amount to almost 800 tonnes. This is a small hill next to the Everest of her son's gift to the climate, born as he as into a world where gases from transport, homes, food and waste were already high and climbing higher.

John Carbone has already amassed three times the emissions his parents had at the same age. But what of his new daughter?

Will her lifetime emissions be even higher than her father's? The Carbones are already helping to reduce Lucy's contribution. From Kate growing much of the baby food, to her mode of transport being a small car rather than the over-sized people carrier of former years, the Carbones are at least making a dent in the Lucy-related greenhouse emissions. They may even be taking some of the heat out of the teenage blame-fests they will face in the future.

What if Lucy were to take climate awareness through into her adult life? What if she lived a low-emissions life: driving a smaller car, avoiding energy wastage in the home, buying more local food, and the like? If we were to stack up such actions, nothing too hair-shirty – no composting toilets or dancing on hilltops with shrubbery in her hair – just choices that fitted with her everyday life, what's the bottom line?

Indulge me with the plotting of another graph (Figure 16; a handful of pie charts was never going to keep the inner scientist at bay). Compare a climate-aware Lucy with a climate-ignorant one: how the emissions build up for these two Alabama girls and how the climate impact they bequeath to the

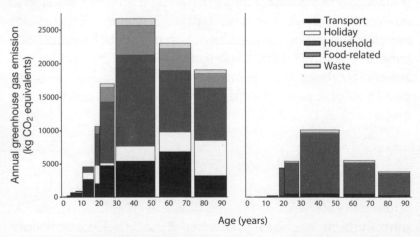

Figure 16 Lifetime emissions of (left) climate-ignorant Lucy and (right) climate-aware Lucy.

world diverges over a lifetime. Let's see where we may all be heading.

On the surface we have two identical Lucy Carbones: one our climate-aware Lucy, born into an increasingly environmentally friendly Carbone household. The other is a Lucy Carbone of climate ignorance. The latter Lucy grows up in the Carbone household where John never did get embarrassed about his big car, where recycling always remained something the left-wing neighbours did, and where Kate's garden was little more than a large patch of concrete and weeds. We'll assume that no magic bullets to tackle climate change have appeared, and that fossil fuels continue to dominate the lives of the two Lucys, just as they have dominated all of ours.

Over at the climate-ignorant Carbone household, things are already stacking up for the baby Lucy. Until she leaves home, many of her emissions are down to the decisions of her parents. Even before she emerged, red-faced and angry this Lucy already had had a raft of important choices made on her behalf. The newly decorated baby bedroom is crammed with enough electronic monitoring devices to put the FBI to shame. Stacked boxes along one wall hold an army of battery-operated activity centres, dolls and singing bears. Nothing, not one pair of baby booties nor one cloth picture book, is second-hand. Everything is brand-spanking new, with the full embodied energy penalty of all this stuff dumped straight onto the tiny shoulders of the ignorant Miss Carbone. This is just the beginning.

Once the Lucys reach 18 months it's straight into the nursery run. While the climate-aware Carbones opt to get to nursery by bike, the ignorant Carbones use their four-wheel drive for these short journeys, so producing an extra 200 kg of greenhouse gas each year. As the years click by, the 'Back to school' signs in the mall once again indicate that, for George, Henry and now Lucy, the days of summer are drawing to a close and it is time to shop for new shoes. With junior school comes the choice between school bus and more travel in the car. The climate-ignorant Carbones stick with the 'safety' of the gas-guzzler to ferry Lucy about, adding another 700 kg of greenhouse gas every year to her already ballooning emissions. Our climate-aware Lucy gets the bus.

School holidays? With her parents having no scruples about clocking up tonnes of emissions flying, jet-setting holidays are the norm for the ignorant Lucy. Though the trips in the early years are marked by some ear-shattering crying, hasty breast feeding in the security queue and, on one regrettable occasion, some projectile vomiting at 9,500 m, the sky-high travel continues. Before she reaches 10 this Lucy has already clocked up 80,000 km and 12 tonnes of greenhouse emissions on jets. Over in the climate-aware Carbone house, their choice to holiday closer to home and avoid the stress and emissions of jet travel mean that the emission gap between the two Lucys has become a chasm.

For the first time, as the Lucys grow through the years at junior school, they become directly responsible for some of their greenhouse gas emissions. Until now our ignorant Lucy could justifiably throw her arms up in the air and flounce around the room at the injustice of being blamed for her high emissions, given that her choice in them has been minimal or nil. Now though, as she goes through junior school, her own choices start to make themselves felt on the climate, these choices reflecting those of her climate-ignorant family.

We're a predictable lot. Despite the rebellion that strikes in our teenage years, we're more than likely to end up voting for the same political party as our parents, eating a similar diet and sounding like them. The effect of our upbringing on our attitudes and behaviour as an adult is powerful, so it's not surprising that the Lucy in the climate-criminal Carbone house has already begun to follow the example set by her parents and brothers. Initially this takes the form of leaving the television, video, and video games console on all the time. This extra energy adds up to an extra 120 kg of greenhouse each year.

Into high school and adolescence, and ignorant Lucy now likes to turn up the heating at home to a level bordering on that of a sauna. Added together, her boost to energy use in the climate-ignorant Carbone house is now equivalent to an extra tonne of greenhouse gas emissions every year.

Soon the two Lucys have made their roller-coaster way through high school. The time has come for the beanbag to be rammed into the car, the dog to be hugged, and the drive to college made. It's time for the Lucys to start college at New York State.

Having left their family homes, the girls now become much more responsible for their personal greenhouse emissions. One of the first and most important decisions they make is their form of transport. Climate-aware Lucy opts for her bike and public transport, while climate-ignorant Lucy chooses to buy and use a sporty black SUV – each year widening the emissions gap by another 6 tonnes. In their dorm rooms, though they may cover their walls in posters and burn enough incense to make the whole floor smell like an aromatherapy shop, their climate change choices remain limited. However, by setting her computer to sleep mode and by replacing the two light bulbs in her bedroom with energy-efficient ones, our climate-aware Lucy shaves another half a tonne off her annual emissions.

The college years fly. Final exams are duly passed and the graduation ball is preserved as a set of embarrassing photos, to be dusted off and examined only when a college-mate runs for president or is found in a clock tower with a sniper rifle. Having finished their studies, they both take on the responsibility of first jobs and homes. They are now directly responsible for the contribution they make to global warming.

We've seen how bad things can get; how, as salaries increase, they can buy bigger cars, take more jet-setting holidays, and eat through more and more electricity as the army of appliances around their house grows. We've also seen that it doesn't have to be like this. Our climate-aware Lucy can have the same standard of living as her ignorant counterpart: she can eat well, stay warm in winter and cool in summer, and travel with ease, all while having a fraction of the impact on our climate and without ever going near a bowl of muesli.

Our grown-up climate-aware Lucy has that small car, the low-mileage food and the efficient boiler. She separates her trash for recycling, avoids air travel and is the scourge of standby power. In those key facets of our lifestyles that affect our climate – our travel, our homes, our food and our backyards – she has slashed her emissions through thinking about it and choosing to do so. Across the board, she meets and exceeds that magic 60% reduction. The cumulative effect across a lifetime? Huge.

By the time the presentation of an engraved carriage clock is made, and plastic cups of fizzy Scottish Chardonnay are passed around, the difference between the emissions glaciers of the two Lucys is stark. At 65 they've had nigh-on 40 years of full control over their emissions and, with each passing year, the extras of ignorant Lucy have mounted up. After a good long retirement of 25 years the two Lucys die and are buried, one in a calico bag, and the other in an airtight testament to conspicuous consumption.

The lifetime impact of our Lucy, for 90 years of climate-aware life in a developed-world country: 595 tonnes of greenhouse gas. For climate-ignorant Lucy: a staggering 1,800 tonnes.

What these two women will have seen by the 2090s we can only imagine. By then, the planet is likely to be substantially warmer and the effects of this severe. Just how much warmer and just how bad the impacts will be depends on how many Lucys of each type there are.

The true graph though, if we could take a real climate audit over the next 90 years, would probably show an even bigger divergence. I've assumed the status quo: the adult climate-ignorant Lucy using energy in the home at the same average rate as her parents, the distances she drives and flies being the same as theirs. In fact, predictions for the year 2025 are for massive rises in average greenhouse emissions from home energy use, food, waste and, the really big one, transport. Emissions from energy use in US homes are predicted to increase by over 25% in the next 20 years, while on the roads we'll be belching out an extra 50% compared with today – over a billion tonnes more every year. For climate-ignorant Lucy therefore, the true story is likely to be much worse. Enough to make anyone turn in a bronze-lined grave.

Add together just some of the emissions-cutting strategies described in the preceding chapters and incorporate them into our lives, just as Lucy Carbone did, and the target of a 60% reduction can be not just hit, but shattered. That the 60% cut is truly a possibility for each of us – something we can do right now, without the pillars of civilisation falling around our ears – represents a fantastic opportunity. Such a reduction is exactly what we need to make if disastrous climate change is to be avoided.

Achieving this 60% cut, one that our politicians can currently only dream of, means that for every person taking action, many hundreds of tonnes of greenhouse gas can be kept out of the atmosphere. Multiply this up for your whole family, your friends, your street, or for all your work colleagues, and you can see exactly how such alterations could literally change the world.

It starts with you and me, at home, at the shops, in the garden and on the way to work. It doesn't have to stop there. Though homes and private transport both play leading roles in global emissions, we have the power to take our new-found awareness of climate change beyond these most personal facets of our lives: we can take it to work.

From what we do with our scrap paper, to whether or not we congratulate John Carbone on his new SUV, this big part of our lives provides a wealth of ways by which we can either ramp up greenhouse emissions or drastically cut them.

So, on to the gleaming corporate tower blocks, to the open-plan offices, the bins stuffed with reams of virgin paper and the lights burning through the day and the night. Let's see just how far-reaching the effects of climate-awareness can be – how one memo from the boss can cut through the energy squandering of dozens of staff, and how just one person in the office remembering to turn the lights out can change our future.

8

lights out

Our climate concerns may be something we'd rather keep quiet about around the water cooler – an embarrassing hobby for the evenings and weekends. Tackling climate change, though, is not some unfashionable activity for people without a life.

Cutting energy wastage and greenhouse gas emissions at work takes us back to the smoking analogy. You've seen the light – smoking (or in this case greenhouse gas emission) is bad for you, for your family, for everyone. You've changed your life-style and kicked the habit, so to speak. You feel good about your enlightenment and want to spread the word. Then, the ultimate test: you go back to work. At 10.15 a.m. on the dot it's the usual cigarette break. Out comes the chocolate bar, the stress ball and the family photo. After a while the smell of smoke reaches your nostrils and, if all that detox worked, it turns your stomach. How can people be allowed to smoke in the same building as non-smokers, exposing you to their smoke and the potentially fatal consequences? You complain. If you're the boss you ban it there and then; if not then the union gets it banned.

In the same way, if you and your family are saving energy and cutting emissions by having low-energy bulbs at home, why should the office be lit like the set of *Ice Cold in Alex*? Why, when at home you pull on a sweater in winter and open the windows in summer, should your office get away with T-shirt temperatures even in the hardest winter snaps and feel like the inside of a refrigerator during the heat of summer? It's not right and, as a climate-aware employee, boss, or even client, by taking a stand against energy wastage in the workplace you can magnify the extent to which you mitigate climate change many-fold. Sure, we can make a big difference at home by being climate-aware, but we can then take that awareness to work, when we go shopping, and maybe most crucially of all, when we step into the polling booth.

On the way home from work, those thousands of brightly lit office windows we pass every evening represent simply massive energy use and wastage. No matter how late at night, those windows will still be shining – and shaming. If you've ever worked in such offices you know just what overheated, overlit, humming computer-fests they are. As an insurance clerk, paid on an hourly rate and always strapped for cash, I'd often chug into the car park on my motorbike well before 7 a.m.

In the streets the traffic would still be a soft murmur compared with the roar of an hour later, and the night security guard on the front desk would be knocking back his eighth and final coffee of the shift. Through the big double doors into the foyer and I'd be hit by wilting heat (particularly in biking leathers) and a hum like an electricity substation.

Up to the open-plan office and the scene was always the same: all the lights on, the computers chuntering away to themselves, and the vending machines humming in unison. Around the room were the familiar array of novelty desk gonks, name cards on computers and photos of last year's office Christmas party. Like some 21st century *Marie Celeste*, the photocopier would be going through another systems check and the drinks machine would be heating up its coffee and cooling down its vat of limeade; again, the only thing missing were the people. From the outside you'd have imagined the building was bursting with activity – from every window on every floor, the lights burned.

Of course, no matter how early I got in there was always somebody in before me. That person in offices the world over – Gareth with the sunken eyes from Loss Assessment in my case – who is always the last to leave at night and the first to arrive in the morning. Maybe Gareth never did go home.

From gnome factories to schools, from butchers to candlestick makers, businesses account for around 40% of greenhouse gas emissions in developed world countries. Put this together with our own residential and transport emissions and you've got the bulk of all developed-world greenhouse gas emissions right there in front of you.

For every square foot of space at your work – about the space your waste bin takes up – the amount of greenhouse gas emitted can be big. At the bottom of our 'emissions at work' league come warehouses. Work in a warehouse and, on average, for every square foot, 4 kg of greenhouse gas will be emitted every year. The bulk of this goes on refrigeration for the stocks in the warehouse and heating, though it may not feel like it, for you and your work mates. A few kilograms of greenhouse gas per square foot may not sound all that much, and compared to some other workplaces, it isn't. But 4 kg per square foot for an average-sized warehouse of, say, 15,000 square feet means 60 tonnes of greenhouse gas a year.

Next come schools. With the heating and cooling of classrooms and halls, the lighting, and the fuel used to produce all those lovely school dinners, the greenhouse emissions rise to 5.5 kg per square foot. The sugar and tartrazine-heavy vending machines that inhabit most refectories can also play a big role, each one being responsible for around two tonnes of greenhouse gas a year, not to mention the dozens of hyperactive kids.

Pop down to the mall for some retail therapy and, from the blast of air-conditioning which greets you as you walk in to the rows of spotlights blazing down on the clothes racks, the retail sector goes up the climate-warming scale again. The lighting-related emissions shoot up to over a quarter of the total, with heating and cooling the other big two. For each square foot, shops emit over 9 kg of greenhouse gas a year: well over 100 kg a year for every rack of sweatshop-stitched tops in this world of dry heat and commercial radio.

Next up the scale come the offices (see Figure 17). Lighting, heating and cooling these big open-plan spaces is again the

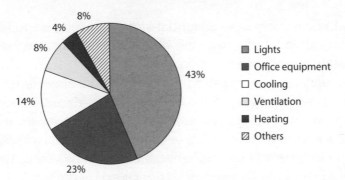

Figure 17 Uses of electricity in the average office.

big draw on power, but all those PCs, photocopiers and vending machines really push things up. That square foot under the office waste bin generates 10 kg of greenhouse gas emissions each year. The office desk, along with its flaking skin of sticky notes, is emitting over 100 kg of greenhouse gas each year. Say you sit in an open-plan office with 19 others, all tucked behind *Dilbert*-style screens, sharing each other's phone calls and drinking stories. On a standard office floor of say 10,000 square feet, the emissions total around 100 tonnes a year. This is over 5 tonnes for you and every one of your colleagues, and equivalent to each of you driving two cars into work everyday instead of one.

I had hoped that the next workplace, my own, would be better. But as university offices and lecture theatres tend to suffer from the same 24 hour lighting, heating and appliance wastage as most offices, the emissions per square foot per year – the damning climate impact of every seat in every lecture theatre – goes up to over 11 kg. This is bad, but we're still only mid-table in the league of work-based warming.

Work in a health centre or hospital? Warm, isn't it? Heating and cooling use up more than half of all the electricity in these places, with all that lighting coming in a close third. Per square foot, the emissions total around 14 kg of greenhouse gas each year. It would be silly to argue that all the energy used in our hospitals is unnecessary. Some good lighting, I imagine, is fairly crucial in the operating theatres; all those machines that go ping also keep people alive; and sick people need to be kept warm. Nevertheless, as places which are likely to see more and more admissions due to climate change-driven heatwaves, disease outbreaks and storm damage, our hospitals are smack in the middle of the two-way street of global warming.

Now to the businesses with really big climate impacts. The next two take things to a whole new level. In second place, with soft lighting out front but a storm of energy use in the back, come restaurants. At over 26 kg of greenhouse gas per square foot, a table for two will clock up around 400 kg of greenhouse gas a year – the same as a year's servings of hand-tied chives. But topping the league for climate impact per square foot are the out-of-town food mile cathedrals that are our supermarkets. Space heating and cooling are pushed unceremoniously into third and fourth places by the huge amounts of energy needed to keep our Chilean grapes and Canadian spring water chilled – a massive 38% of total electricity use in this biggest of business energy users. Lighting for all those lovely packets of biscuits comes in as the second biggest drain on energy. In total, every square foot of your local grocery store accounts for over 30 kg of greenhouse gas. Walk down just one aisle and your trolley will have passed through an area responsible for 12 tonnes of greenhouse gas each year – as much as a large four-wheel drive.

Back when I was researching for the chapter on food and its greenhouse emissions, I volunteered to take myself and my daughter to the local supermarket, do the week's shopping and check out the array of food-mile-heavy produce on offer. This cunning plan went somewhat awry when said daughter took exception to my wanting to go down any aisle other than the one with cakes in it. My scribblings on the country of origin were therefore rather frantic. I did, though, manage to record the contents of one whole stand of luxury groceries. This stand was about 2.5 m long and 1 m deep, and on its shelves were a selection of plastic-packaged delights from all around the world, including runner beans from Thailand, baby carrots from South Africa and blueberries from New Zealand. But, sadly, nothing that could keep a two-year-old interested for more than the time it takes to ram a baby carrot up your nose.

My intention at the time was to calculate the contribution to global warming of this one short section of shelving. Having got carried away with hand-tied chives and lead grapes earlier, this somehow slipped by. Now, though, we've got the climate impact not only of the food miles, but also the shelves the food is sitting on.

Add together all the packets of blueberries, baby carrots and other jet-setting produce on this stand, assume just 20 packets of each type are bought every week, and the greenhouse gas emissions for this one stand take on mammoth proportions. The food mile emissions add up to over 40 tonnes a fortnight, and over 1,000 tonnes in a year. Combine these emissions with the energy used to keep all those pocket-sized packets brightly lit and cool, and in a year we've got emissions equivalent to driving 50 gas-guzzling SUVs all the way around the world.

To make real reductions in greenhouse emissions at your work-place can seem rather tricky. For one thing there's no clear monetary incentive for your colleagues, and for another, tell-ing your boss she should turn down the heating and put on a sweater may go down about as well as when you try it with your teenage kids. Just as at home though, many of the smallest changes can have big effects.

Let's start with that over-lit insurance office of my biker days. The office has 20 big lighting units, each with four-lamp 4 ft fluorescent lights. They are on for 100 hours a week, 365 days a year. Now, before we start making Gareth sit in the dark or shut down the light show which is our office block at night, what if we just changed these fittings for more efficient bulbs?

By swapping over to more efficient bulbs we'd each year cut our office energy use by around 3,700 kilowatt hours, save the company over $1,000 dollars in energy costs and cut 2.3 tonnes off annual greenhouse gas emissions. We can go much further though. The current lights produce more than a third more light than that required for the work that's done, so, install enough energy-efficient lighting to hit the ideal lighting level for the office and we'd save almost 7,000 kilowatt hours – over 4 tonnes of greenhouse gas a year.

Always-in Gareth may not like it, but how about the lights being turned off when they're not needed? His frighteningly tidy area of the office aside, if the other lights were turned off outside of work hours then we'd go from 100 hours of lighting per week to just 50 hours, halving energy use and greenhouse gas emissions in a stroke.

Turning the lights out really gets to the crux of saving energy at work. So often the fact that it's not our own energy bill, and that someone else can always do it, means the lights in offices and factories are just left to burn. Aside from relying on you and me to break this cycle and flick that switch, there is the option of occupancy detectors. These mean that the lights go on when Gareth first skulks through the maze of desks at some ungodly hour in the morning, and turn themselves off again at night when he's finally had enough of trying to complete Minesweeper.

Imagine them as standard in schools, reacting to the constant incoming and outgoing tide of classes throughout the school day and so saving energy hand over fist. In the office they'll save up to 50%, and in the bathrooms at work up to 75% (depending on how much coffee everyone drinks). The same big cuts hold true for meeting and conference rooms (up to 65%), corridors (40%) and in the cavernous space of a warehouse (up to 75%).

After lighting, one of the most conspicuous consumers of energy at work is office equipment. And, just as for lighting, the savings to be made simply through turning things off are big. The desktop computer, that digital window we spend so many hours staring at, is the biggest player here. PCs and their monitors have become so ubiquitous that their always being on and draining power is something that has become part of the background scenery of every office. Offices across the world go through each night to the hum of countless PCs, fans and hard-drives whirring and clicking, 'The Matrix Has You' screensavers whirling around their screens. Turn them off at night and for every PC without a hum 675 kilowatt hours of electricity are avoided each year – over 400 kg of greenhouse gas emissions saved just for hitting that switch. The photocopier over by the wall? Turn that monster off each night and, as well as helping to prevent unsavoury photocopied body parts mysteriously appearing in the boss's in-tray, you'll cut a buxom 4 tonnes off annual emissions.

While we're on the subject of turning things off, a brief mention of screensavers. These scrolling messages, virtual fish tanks and *Star Trek* scenes might look rather jolly for the first two seconds they're in operation, but they do nothing for saving energy. Instead, they keep the monitor and the PC eating energy while you're off at lunch, rather than allowing them to power down.

Just as some fridges can honk out much more greenhouse gas than their more energy-efficient counterparts, so the printer you have in your office, the photocopier you order for your floor, and the computers on all the desks, come in climate-friendly as well as unfriendly versions. Replacing defunct office equipment with Energy Star equipment together with turning those switches off at night, takes the greenhouse gas cuts possible up to almost three quarters – around half a tonne of greenhouse gas for every PC and well over five tonnes for the copier. Want a new printer or fax machine for the office? Give laser printers a wide berth – these emit about half a tonne of greenhouse gas a year compared to just 20 kg for inkjets.

The same goes for the office fax, scanner and printer. Opt for energy-efficient versions and, during the times when the printer isn't spewing out a 100-page mobile phone user manual, and the fax machine isn't delivering offers of the ultimate in lard-only diets, they'll power down and so cut energy wastage and emissions by up to 50%.

With the average vending machine being responsible for around 2 tonnes of greenhouse gas a year, there are big savings to be made by opting for energy-efficient models. Using features such as proximity detectors – the vending machine only goes to full power when approached by caffeine- or chocolate-starved customers – can halve its emissions. In 2000 the schools of Moscow, Idaho, had 20 of their vending machines upgraded to increase energy efficiency. This was so successful that a further 250 vending machines in offices and public buildings throughout the community were also upgraded. The combined energy savings of these upgrades meant a cut of around 200 tonnes of greenhouse gas every year.

For the energy-intensive floor space in our supermarkets, restaurants and the like, having more energy-efficient appliances becomes even more important. As with vending machines, greenhouse gas emissions due to food store refrigerators and freezers can also be halved by opting for more energy-efficient models and keeping them well maintained.

Outside of direct energy use at work, there's one other major burden on the climate that's everywhere you look. Across our desks, filling the cabinets, in piles on shelves, and pouring out of the waste bins – paper.

Back in the early 1990s, as the Internet was just starting to spread its web across the world, personal computers like the one I'm using now were already eating their way through around 115 billion sheets of paper. Today, despite broadband connections and email as standard in most offices, paper use has reached staggering proportions. In the USA alone, laser printers are now getting through over a trillion pages each year and paper consumption continues to rise by 20% annually. Each of us uses an average of 100 sheets of paper a day, the bulk of this ending up in the bin. The paper used by a typical office worker over a year carries with it embodied energy equivalent to more than 100 kg of greenhouse gas. That's before it even gets to the landfill site.

While researching studies of paper use in offices and the effect of the Internet, what did I do? I used my PC, found some online papers and printed them out to read them (double-sided of course). Email, I had always assumed, was bound to have cut our paper use in the office. Surely, we don't send as many letters and memos as a result? I was wrong. On average, the use of email in an organisation actually increases paper use by 40 per cent.

The ways around having the equivalent of a finely sliced forest scattered around your workplace are many and, like flicking the light switch, take no more than a moment's thought. Need to make 500 copies of the interim report on expanding air-conditioning sales? Hit the double-sided copying button and save.

If you hold the precious keys to the stationery cupboard or have a hand in the paper that the office uses, providing recycled paper can make a real dent in the climate impact of your workplace. This doesn't mean a choice between grey and lumpy sheets and bleached-white paper. Much recycled paper can be impossible to distinguish from more energy-intensive virgin paper. Across a whole company the savings, energy-wise and climate-wise, of using it can be huge: the double-sided feature on the copier plus recycled paper saves two whole trees and a tonne of greenhouse gas for every 100 reams of paper used at work. To complete the recycling cycle, as it were, recycling office paper rather than sending it to the local landfill has the same straightforward climate benefits as we saw for our Sunday newspapers – around 2 kg of greenhouse gas saved for every kilogram recycled.

Providing recycling for cans, glass and the rest at work can also reap big dividends. Take the mass of aluminium cans resulting from a hundred desktop lunches every day, five days a week, 48 weeks a year in an office building – let's say a tonne of aluminium cans a year. To make these cans from virgin materials means the emission of about 20 tonnes of greenhouse gas; recycle this lot and next year's cans will be responsible for the emission of just 3 tonnes of greenhouse gas.

Our monster-size copying machine is a great example of somewhere that a range of energy savings and greenhouse gas cuts come together. Over its seven-year life span (barring buttock-related accidents), your standard copier will eat its way through electricity, paper and toner cartridges equivalent to 80 tonnes of greenhouse gas – the annual emissions of two average households for this one copier. A combination of an energy-efficient model, double-sided copying, and the use of recycled paper and toner cartridges means a cut of up to 75% in energy use, cost and greenhouse gas emissions, as well as avoiding the dicing up of 50 trees.

Work from home, like our tele-worker in his log cabin retreat, and the same sorts of savings apply. For an average home office with the standard printer, fax, small photocopier and PC, annual energy use will mean almost two tonnes of greenhouse gas emissions. Switch to energy-efficient equipment and settings and turn things off when they're not being used, and you'll get a cut in greenhouse gas emissions of 60% – over a tonne a year. Likewise with paper use at home: print on both sides and recycle the scrap stuff, and the annual greenhouse gas emissions will fall from about half a tonne to less than 50 kilograms.

One of the biggest energy hogs and greenhouse gas emitters at work tends to be heating and cooling. Before we touch the thermostat, simply opting for energy-efficient appliances can help a lot here too. Less wasted energy means less heat emitted, and so less need for the air-conditioning to be up full blast. For the average office, just having more energy-efficient equipment will cut the electricity needed by the air-conditioning by a third. After that, you can of course reach for the thermostat, but this is where our own control can often start to slip.

If your office is too hot you may be able to turn down the radiator, but in most offices the temperature is set by a control panel more jealously guarded than even the stationery cupboard key. So, rather than risk the wrath of Barbara, she of the poor circulation and undisputed Queen of the thermostat, we set up a forest of fans in our offices, blowing the over-heated air around the office, eating through yet more power and occasionally kicking up a virgin-paper storm. In open-plan work environments it's up the boss to ensure that the heating and cooling of the building isn't a non-stop fight between electric heating and cooling. If they do take such action – providing Barbara with her own heater and allowing everyone else to cool down a bit – the energy savings and emissions savings can be big. Exactly how big? Let's visit John Carbone one last time.

John's office is just like the dozens of others clustered around the south side of Greenville, Alabama. Today he's driving into work early with his fellow carpoolers to avoid the hundreds of other office workers who jam the roads leading into town each morning. He quickly finds a parking space amongst the acres of soon-to-be crammed car park, and heads towards his glass-fronted building. Even at this early hour it is ablaze with the lights of 200 offices. Within this hub of Alabama insurance there reside 200 PCs, 20 networked laser printers, five fax machines, five scanners and five giant-sized photocopiers, all munching through electricity round the clock. Determined to do something about the masses of energy he can see going to waste, and now, as Area Manager, being in a position to do something about it, John Carbone is set to slash energy use, costs and the greenhouse emissions of his workplace.

That same week, occupancy detectors are installed for the lighting system, immediately cutting the energy use for lighting by half and greenhouse gas emissions by 19 tonnes a year. Buoyed by the obvious savings in the next electricity bill, the process of ensuring that PCs, photocopiers and all the other energy-hungry devices are set to their energy-saving modes and turned off at night begins. With a few choice people on board, each known for their abilities in organising collections for leaving parties, John quickly succeeds in cutting energy waste from all the office equipment by over 60% – over 100 tonnes of greenhouse gas emissions a year across the building. With this same team of office matriarchs, recycling becomes a matter of course on all floors. The bins are quickly transformed from their standard state – overflowing with virgin paper – to scrupulously managed recycling areas for paper, cans and bottles.

When it comes to renewing the lease on the 10 company cars, it's no more of those 3.5 litre gas-guzzling saloons for John Carbone and his staff. Instead, at a much reduced cost, the sales force now drive dual-fuel hatchbacks. This simple move cuts the fleet greenhouse gas emissions by over 50 tonnes a year. With new bike racks, shower facilities for the sweaty few who do brave the car-clogged roads, and a designated free parking zone for car pool users, John is well on his way to creating a truly climate-aware workplace.

Through his actions John has cut the greenhouse gas emissions of his office building by over 150 tonnes a year. This is the equivalent of taking a dozen cars off the road permanently, or of slicing over a tonne off the individual emissions of everyone working in the building. John Carbone's boss in the San Francisco head office will also be very pleased. He's cut the electricity bill by over $30,000 a year. Together with the fuel and hire savings on the company cars, and the reduction in paper costs around the office, John's earned the office a hefty Christmas bonus and set the standard for the other regional offices to follow.

As we go up the power ladder at work, the potential for such sweeping reductions in workplace emissions grows and grows. As well as promoting cycling to work and switching to more fuel-efficient company cars, options like choosing a renewable electricity supplier, providing financial support for office recycling schemes, even ensuring climate-aware building design, become possible.

The effects of such workplace energy initiatives are already being felt. In the USA, the Energy Star Programme has expanded to include more than 13,000 organisations and small businesses. Since its inception in 1992, it is estimated that Energy Star has helped save around 55 billion kilowatt hours of energy use by US businesses – the equivalent of over 30 million tonnes of greenhouse gas.

In the UK, my research funding body, the Natural Environment Research Council, now has a Green Audit each year, with the aim of reducing environmental damage across the board. Such organisation-wide greening, from cutting energy use to reducing paper wastage, is happening more and more frequently as a straight money-saving approach, through a wider concern for public image, or maybe even simple altruistic intentions (well, you never know).

The Guardian newspaper now has an annual audit by the Carbon Trust. Initially, this threw up some unpleasant facts. Its offices were producing 418 kg of greenhouse gas per square metre per year (even worse than a supermarket). It is now setting out to cut these emissions to the so-called Good Practice level of 95 kg per square metre. This will mean a cut for the newspaper as a whole of nearly 80% – 2,800 tonnes of greenhouse gas a year. This is all great news for a paper keen to reduce its damage to the environment, and for its bank balance. The potential energy cost savings amount to £120,000 a year.

Not to be left out, the World Wide Fund for Nature has got companies around the world to commit to cutting emissions through its Climate Saver Program. In short, from the first recycling trays appearing in our offices, right up to major corporations having green audits, climate change action is coming to work.

While we're on the subject of the ripple effect of awareness, it's worth mentioning how our attitudes can create changes right through to the very top. Politicians are all too aware of the need to either react to public opinion or face being out of power at the next election. So too are the owners of shopping malls, supermarkets, and appliance manufacturers. Get enough people complaining about digital TV boxes that can't be turned off and the manufacturers will soon start ensuring that there is an OFF switch as standard. If sales of high-efficiency light bulbs at that big blue shop from Sweden take

off, then your local hardware store will suddenly sprout a special offer on these 'new' cost-saving bulbs.

Our bottom-up power to affect the decisions of everyone
from supermarket chains to Presidents and Prime Ministers
can't be overstated. These people of great power are still just
that, people. They have homes, kids and friends who are just as
threatened by climate change as we are. In the UK even the
Queen recently opened a conference on climate change in
Germany, expressing her concern over the issue and her intention to cut the Royal climate impact.

From a political perspective, the promotion and subsequent
incorporation of individual-level cuts into national greenhouse
gas budgets seems an obvious tack to take. Recent years have
certainly seen increasing government interest in this area. The UK
government is promoting domestic energy efficiency via better
information, financial incentives and tighter regulations. The prediction is that implementation of these strategies will cut UK
carbon emissions by around 5 million tonnes by 2010, all without
having to upset any friends in big business. Other governments
are even more proactive in the promotion of emissions cuts at the
individual level. The Australian Greenhouse Office, for instance,
funds Cool Communities, a programme which not only provides
information on how individuals might reduce their emissions, but
also provides grants for communities to implement these household greenhouse gas reduction strategies.

No definitive figures exist for the cost of reductions via the
increased public awareness route, but there's little doubt that
this option has huge potential in industrialised countries. For
every million people with the old SUV-driving John Carbone
lifestyle that switch to his new climate-aware lifestyle, the
annual reduction in greenhouse gas emissions would be more
than 10 million tonnes.

Over the next few years, each of us will be bombarded with
energy-saving leaflets, recycling schemes and public aware-

ness ads. There will be more climate tax carrots and sticks, more harrowing TV news coverage of the impacts of rapid climate change, and our kids will get increasingly cross with us. In the end though, it's our choice.

The greenhouse emissions from the smoky power station down the road can appear so great as to make our own actions seem meaningless. But much of the electricity it produces is then wasted on lighting empty offices, standby power in our homes and the rest. In the USA alone, energy wastage due to standby power requires the equivalent of 26 good-sized power stations, all just to keep the hum going. In fact, it's the energy and fossil fuel we use every day that dominates emissions and so represents the key to tackling human-made climate change.

About a quarter of all developed-world emissions come from our homes, another quarter from our transport, and most of the rest from our places of work (Figure 18). Now, imagine taking the chunks out of this emission pie that we've seen are possible: a 60% slice out of transport emissions; wiping away up to three-quarters of those residential emissions; and then going on to tackle the huge wedge that is the emissions from business. From the car we choose to drive to whether we turn the office PC off at night, it's these things which will decide the severity of climate change in the 21st century and beyond.

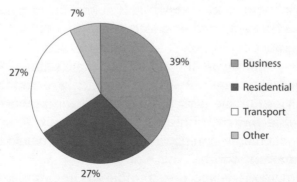

Figure 18 Carbon dioxide emissions by end user (UK, 2001).

If you combine this potential for emission cuts at an individual level with increased awareness, you've got to be optimistic. Our children and grandchildren don't *have* to face the full horrors of a high-emissions scenario. We don't *have* to go down as one of the most selfish and destructive generations in human history; this age of human control of the planet – this Anthropocene – doesn't have to consume itself and everything else.

I've talked aplenty about the magic 60% target – the reduction in emissions that the scientific community believes is necessary if we are to avoid the most devastating effects of climate change. But what, as far as the effects on our homes and holidays, our livelihoods and our health, will such cuts in our greenhouse gas emissions actually achieve?

Predictions of global warming rely on so-called Emissions Scenarios. These are the pathways humankind must choose between. Along each of these paths various changes in climate await. There are many scenarios, together covering the changes in population growth, economic development and greenhouse gas mitigation efforts that we may see over the next 100 years.

The most familiar of the paths ahead of us, the one that means, in theory, our lifestyles can go on relatively unchanged, leads to many of the most severe effects. In the world of climate science this path is part of the so-called A1 Story Line, a family of possible Emissions Scenarios sharing a future of rapid economic growth. The high-emissions scenario, called A1F1 (climate modelling doesn't go in for the most evocative names), represents a world where the high rate of international economic growth is driven by fossil fuel use. It's a world where stockbrokers, defence conglomerates and oil magnates remain the dominant powers, where the 'make money at all costs' ethic continues apace, doing wonders for the sales of red braces but pushing the rate of greenhouse gas emissions up and up. In the A1F1 story line we have the sleeping emissions giants of China and India developing in the same greenhouse gas-intensive way that we in the West have done during the 19th and 20th centuries.

In this chilling story, the world's population peaks by 2050 and then starts to decline. At the same time this scenario assumes the rapid development and use of more efficient technologies. Down this path, greenhouse gas emissions finally start to level off somewhere between 2050 and 2080. By this time though, carbon dioxide concentrations in the atmosphere will be more than double what they are today, and triple what they were before the Industrial Revolution – somewhere up around 800 parts per million (ppm). By following this long-burn path we will have loaded the climate change dice with over 2,000 billion tonnes of extra carbon, and so drastically shortened the odds on catastrophic climate change.

Round that first bend of the high-emissions path, only just out of sight for many of us, there lurks real danger. Under a high-emissions scenario, the UK, for example, faces an increase in the frequency of extreme sea levels, storm surges and flooding along the east coast of between 10 and 20 times what it is today. By opting for the high-emissions path, the brutally straightforward impact tables produced by the Intergovernmental Panel on Climate Change (IPCC) – their projections of disease, famine, flooding and death – are made flesh.

For the range of futures currently predicted, the IPCC has identified five Reasons for Concern categories (Figure 19). For each Concern there is a long box, at the left-hand end of which we have the situation today. As we look along to the right we

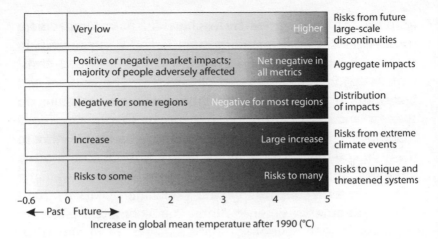

Figure 19 Reasons for Concern.

move into the future, ending up at 2100. As you'll have noticed, these five boxes all have one thing in common: they get a deeper shade as we move towards the right. With a high emissions scenario and a global temperature rise of 5 °C, we get the deepest of shadow – not good.

The top box is for 'Risks from future large-scale discontinuities' and goes from 'Very low' now, to the rather non-specific 'Higher'. This doesn't sound all that bad until you find out what it means. These are the risks of our greenhouse gas emissions having severe impacts right around the world: things like collapse of the West Antarctic ice sheet – something which would push up sea levels by around 6 metres and lead to death and destruction on a global scale. Then there's the shutdown of the North Atlantic circulation and the ushering in of Arctic conditions to huge swathes of North America and Europe.

Below this Hollywood-fodder box the Reasons for Concern may get less dramatic but the boxes quickly grow an ever-deeper shade. First there are the boxes for the economic costs and geographical distribution of global warming. These go from today's situation, where we have 'Negative impacts for some regions' to the spread of impacts and financial costs to every nation, rich or poor, summed up as 'Negative for most regions' and 'Net negative in all metrics'. The bottom two boxes, those representing the 'Risks to unique and threatened systems' and of 'Risks from extreme climate events' are already shading given the warming we're experiencing today. Move to the right and through time, and a high-emissions scenario eats up more and more unique ecosystems and sees a large increase in the frequency of extreme climatic events.

The high-emissions scenario then, the A1F1 future of booming fossil fuel-fed economies and emissions, is a path paved with riches but leading to ruin. Thankfully, at the climate change crossroads at which we now find ourselves, there are other possibilities. Propping up the league of Emissions Scenarios used by the IPCC comes the Low Emissions family and its B1 Story Line. In this B1 future we have the world's population increasing up to the middle of the century, just as for the high-emissions story line, but in this future the world's population has a high level of environmental awareness. This is combined with a rapid change to a more information-based world economy, ensuring more sustainable development of expanding economies such as China. There's also a big and sustained drop in material intensity around the world – simply put, a big cut in the amount of stuff we make, use and throw away. Under the B1 scenario, carbon dioxide concentrations rise to 520 ppm by 2080 and global temperatures increase by about 2 °C. This 'best case scenario' may still represent a drastic increase in global temperature, but at least it keeps us out of the dark ends of the Reasons for Concern boxes.

Of all the Emissions Scenarios, none sees a cut in global emissions until the middle of this century and only one story line, good old B1, then sees a decrease to a level below that which we are emitting today. On the basis of these predictions then, all the paths open to us, from the high-emissions A1 scenario to the low-emissions B1, mean large-scale losses of jobs, homes and lives. At our climate crossroads though, there is one more path.

It's a path the climate modellers might be forgiven for over-looking. It doesn't rely on international climate change initiatives, on the development of 'silver bullet' zero-emission technologies, or even a falling global population, to cut global greenhouse gas emissions. Rather, the direction of this emissions path is nudged downwards by individual awareness and action – very tricky to predict but something with huge potential. The increased environmental awareness so important for the B1 scenario is expected to come from clear evidence of rapid climate change – our collective horror at the famines, floods and disease epidemics that climate change will bring. What though, would the future hold if we acted now? If, instead of waiting for climate change to hit us, we hit it?

This is no happy-clappy scenario of zero cars, planes or electric lighting. Rather, it's a real possibility. It's the sum of the world's Carbones doing their bit and more; it's Barbara and millions like her cutting energy use at the office; hell, it's even the world's Gareths turning off their PCs and cycling home from work. As individuals they are no more than grains of sand in the path of the climate change glacier; together they can radically change that path.

Extrapolate the cuts that we've seen are possible at an individual level to a global scale and the reduction in global greenhouse gas emissions would be like six Kyoto Protocols stacked on top of one another: no loopholes and backsliding, just real and significant cuts. If the emissions that each of us is responsible for, from our homes, transport and jobs, dive in this way, then we get a new scenario – let's stick with IPCC convention and call it 'C1'. This global future sees a carbon dioxide concentration in 2080 of below 500 ppm, avoids the 2 °C global temperature increase, and saves the world from the worst of the potential climate impacts.

This 'C1' path is our chance to make a real difference, a path open to us all. We're standing at this climate change crossroads right now.

Choose.

Figure 20 Flooded sign.

references and resources

Chapter 1

Australian Greenhouse Office (2001) *Global Warming: Cool it!* http://www.greenhouse.gov.au/gwci/

Central Alabama Electric Cooperative. *Residential Energy Calculator.* http://touchstoneenergyhome.apogee.net/index.asp?id=centralal

Department of the Environment and Transport in the Regions (DETR). *Vehicle Certification Agency Emissions Database.* http://www.vcacarfueldata.org.uk/search_form_basic.asp

Energy Information Administration. Average Electricity Emission Factors by State and Region, USA. http://www.eia.doe.gov/oiaf/1605/e-factor.html

Friedland, A. J., Gerngross, T. U. and Howarth, R. B. (2003) Personal decisions and their impacts on energy use and the environment. *Environmental Science and Policy*, **6**, 175–9.

Greenhouse Gas Online. http://www.ghgonline.org/

Intergovernmental Panel on Climate Change (IPCC) (1999) *Air Transport Operations and Relation to Emissions. Aviation and Global Atmosphere.* Cambridge University Press, Cambridge.

Liverman, D. M. and O'Brien, K. L. (1991) Global warming and climate change in Mexico. *Global Environmental Change*, **1**(5), 351–64.

Reay, D. S. (2002) Costing climate change. *Philosophical Transactions of the Royal Society Series A*, **360**, 2947–61.

United Nations Convention on Climate Change (UNFCCC) (2003) *Caring for Climate: a Guide to the Climate Change Convention and the Kyoto Protocol.* http://unfccc.int/resource/docs/publications/caring_en.pdf

US Environmental Protection Agency. *Global warming.* http://yosemite.epa.gov/oar/globalwarming.nsf/content/index.html

US Environmental Protection Agency. *Greenhouse gas emissions from management of selected materials in municipal solid waste.* http://www.epa.gov/epaoswer/non-hw/muncpl/ghg/chapter4.pdf

US Environmental Protection Agency. *Solid waste management and greenhouse gases: a life-cycle assessment of emissions and sinks.* http://www.epa.gov/epaoswer/non-hw/muncpl/ghg/ghg.htm

WasteWatch. *WasteOnline: In depth information on waste.* http://www.wasteonline.org.uk/

Chapter 2

Argonne National Laboratory. Center for Transportation Research http://www.transportation.anl.gov/

Australian Greenhouse Office. Fuel Consumption guide: 10 top tips for fuel efficient driving. http://www.greenhouse.gov.au/fuellabel/costs.html#tips

Bureau of Transportation Statistics, US (2003) *America on the Go*. National Household Travel Survey, USA. http://www.bts.gov/programs/national_household_travel_survey/

Caldwell, H. *et al.* (2002) Potential impacts of climate change on freight transport. *The Potential Impacts of Climate Change on Transportation Workshop 2002*. http://climate.volpe.dot.gov/workshop1002/caldwell.pdf

Cooper, J., Ryley, T., Smyth A. and Granzow, E. (2001) Energy use and transport correlation linking personal and travel related energy uses to the urban structure. *Environmental Science and Policy*, **4**, 307–18.

Elsasser, H. and Burki, R. (2002) Climate change as a threat to tourism in the Alps. *Climate Research*, **20**(3), 253–7.

Energy Information Administration, US (1996) *Alternatives to Traditional Transportation Fuels 1994*. Volume 2: *Greenhouse Gas Emissions*. http://www.eia.doe.gov/cneaf/pubs_html/attf94_v2/exec.html

European Environment Agency. *Vehicle Occupancy Rates*. http://themes.eea.eu.int/Sectors_and_activities/transport/indicators/technology/TERM29%2C2002/TERM_2002_29_EU_Occupancy_rates_of_passenger_vehicles.pdf

Fourth Virtual Conference on Genomics and Bioinformatics. 2004. http://www.virtualgenomics.org/conference_2004.htm

Friends of the Earth (FOE). *Why travelling by rail is better for the environment*. http://www.foe.co.uk/pubsinfo/briefings/html/20011012100132.html

FuelEconomy.gov. US Department of Energy. http://www.fueleconomy.gov/

Greene, D. L. and Schafer, A. (2003) *Reducing Greenhouse Gas Emissions from US Transportation*. Report for the Pew Center on Global Climate Change. http://www.pewclimate.org/global-warming-in-depth/all_reports/reduce_ghg_from_transportation/index.cfm

Intergovernmental Panel on Climate Change (IPCC) (2000) *Comparison of Carbon Dioxide Emissions from Different Forms of Passenger Transport. Aviation and the Global Atmosphere*. Cambridge University

Press, Cambridge. http://www.grida.no/climate/ipcc/aviation/126.htm

IRFD World Forum on Small Island Developing States. *Challenges, Prospects and International Cooperation for Sustainable Development.* http://irfd.org/events/wfsids/vc.htm

The Leading Edge: Second National Conference for the Stewardship and Conservation Community in Canada, 2003. http://www.stewardship2003.ca/

Lise, W. and Tol, R. S. J. (2002) Impact of climate on tourist demand. *Climatic Change,* **55**(4), 429–49.

Macedo, I. D. (1998) Greenhouse gas emissions and energy balances in bio-ethanol production and utilization in Brazil. *Biomass and Bioenergy,* **14**(1), 77–81.

Maddison, D. (2001) In search of warmer climates? The impact of climate change on flows of British tourists. *Climatic Change,* **49**(1–2), 193–208.

McCleese, D. L. and LaPuma, P. T. (2002) Using Monte Carlo simulation in life cycle assessment for electric and internal combustion vehicles. *International Journal of Life Cycle Assessment,* **7**(4), 230–6.

National Greenhouse Gas Inventory, Australia (2002) *Energy: Transport. 2002 Inventory and Trends.* http://www.greenhouse.gov.au/inventory/2002/facts/pubs/02.pdf

Prodmore, A., Bristow, A., May, T. and Tight, M. (2003) Climate change, impacts, future scenarios and the role of transport. *Working Paper 33,* Tyndall Centre for Climate Change Research, UK. http://www.tyndall.ac.uk/publications/working_papers/wp33.pdf

Randall, F.J., Driscoll, W., Lee, E. and Lindsay, C. (1998) *Greenhouse gas emission factors for management of selected materials in municipal solid waste.* US Environmental Protection Agency. http://yosemite.epa.gov/OAR/globalwarming.nsf/UniqueKeyLookup/SHSU5BVP7P/%24File/r99fina.pdf

Reay, D.S. (2003) Virtual solution to carbon cost of conferences. *Nature,* **424**, 251.

Reay, D.S. (2004) Flying in the face of the climate change convention. *Atmospheric Environment,* **38**, 793–4. http://www.ghgonline.org/flyingaea.pdf

Root, A., Boardman, B. and Fielding, W. J. (1996) *SMART: The Costs of Rural Travel.* Energy and Environment Programme, Environmental

Change Unit, University of Oxford, UK. http://www.eci.ox.ac.uk/pdfdownload/smartreport.pdf

Sausen, R. and Schumann, U. (2000) Estimates of the climate response to aircraft CO_2 and NOx emissions scenarios. *Climate Change*, **44**(1–2), 27–58.

Scott, B. M. and Plug, L. J. (2003) CO_2 emissions from air travel by AGU and ESA conference attendees. *EOS Transactions*, Fall 2003 Meeting of American Geophysical Union. http://surface.earthsciences.dal.ca/publications/abstracts/scottplug_agu2003.pdf

Shackley, S. *et al.* (2002) *Low carbon spaces area-based carbon emission reduction: a scoping study.* Sustainable Development Commission. http://www.tyndall.ac.uk/research/theme2/final_reports/sdc_final_report.pdf

Strategic Rail Authority, UK. *The way forward for Britain's railway: relative (CO_2) emissions for different modes of transport.* http://www.sra.gov.uk/publications/general/general_The_Strategic_Plan_2002/strategic_planthe_way_forward.pdf

Thomson, S. (2001) *The impacts of climate change: implications for the DETR.* Report for the Department of the Environment, Transport and the Regions by the In House Policy Consultancy Unit, UK. http://www.defra.gov.uk/environment/climatechange/impacts/pdf/impacts.pdf

Thomson, S. (2003) The impacts of climate change: Implications for Defra. Report for the Department of the Environment, Food and Rural Affairs by the In House Policy Consultancy Unit, UK. http://www.defra.gov.uk/environment/climatechange/impacts2/pdf/ccimpacts_defra.pdf

Toohey, R. (2001) *Travelling beyond boundaries? Catch a bus!: a rural perspective on public transport.* Conference Papers. Institute of Public Works Engineering, Australia. http://www.ipwea.org.au/papers/download/Royce%20Toohey.doc

Tyndall Centre for Climate Change Research. *Carbon emissions from transport: relative (CO_2) emissions for different modes of transport.* http://www.tyndall.ac.uk/research/info_for_researchers/emissions.pdf

UK Department for Transport. *Energy and environment: emissions for road vehicles (per vehicle kilometre) in urban conditions.* http://www.dft.gov.uk/stellent/groups/dft_transstats/documents/page/dft_transstats_032073.pdf

UK Department of Transport (2003) *GB National Travel Survey.* Personal travel factsheets. http://www.dft.gov.uk/stellent/groups/dft_control/documents/contentservertemplate/dft_index.hcst?n=7223&l=3

UK National Atmospheric Emissions Inventory. *Road Transport.* http://www.aeat.co.uk/netcen/airqual/naei/annreport/annrep98/app1_29.html

US Environmental Protection Agency. *On the Road.* http://yosemite.epa.gov/OAR/globalwarming.nsf/content/EmissionsIndividualOntheRoad.html

US National Biodiesel Board. http://www.biodiesel.org/

Vehicle Certification Agency (VCA). *Car fuel data.* UK Department of Transport. http://www.vcacarfueldata.org.uk/

Wang, M., Saricks, C. and Santini, D. (1999) *Effects of fuel ethanol use on fuel-cycle energy and greenhouse gas emissions.* Center for Transportation Research. Argonne National Laboratory. http://www.transportation.anl.gov/pdfs/TA/58.pdf

Wang, M., Saricks, C. and Wu, M. (1997) *Fuel-cycle fossil energy use and greenhouse gas emissions of fuel ethanol produced from US Midwest corn.* Report for Illinois Department of Commerce and Community Affairs. Center for Transportation Research. Argonne National Laboratory. http://www.transportation.anl.gov/pdfs/TA/141.pdf

Wasteline. WasteOnline UK. *End-of-life vehicles.* http://www.wasteonline.org.uk/resources/InformationSheets/vehicle.htm

Yang, M. (2002) *Climate change and GHGs from urban transport.* Asian Development Bank. Transport, Planning, Demand Management and Air Quality Workshop. Manila, Philippines. Document 10b. http://www.adb.org/Documents/Events/2002/RETA5937/Manila/downloads/cw_10B_mingyang.pdf

Chapter 3

Australian Consumers' Association. *Standby Wattage – Standby Wastage.* http://www.choice.com.au/viewArticle.aspx?id=102226&catId=100447&tid=100008&p=1

Australian Greenhouse Office. *Embodied energy.* http://www.greenhouse.gov.au/yourhome/technical/fs31.htm

Australian Greenhouse Office. *Strategic study of household energy and greenhouse issues.* Prepared by Sustainable Solutions Pty Ltd, June 1998. http://www.greenhouse.gov.au/coolcommunities/strategic/

Australian Institute of Energy. *Energy value and greenhouse emission factor of selected fuels.* http://www.aie.org.au/melb/material/resource/fuels.htm

California Energy Commission. *Consumer tips for appliances.* http://www.consumerenergycenter.org/homeandwork/homes/inside/appliances/

Coley, D. A., Goodliffe, E. and Macdiarmid, J. (1998) The embodied energy of food: the role of diet. *Energy Policy,* **26**(6), 455–9.

Community Carbon Reduction Project (CRED), UK. http://www.cred-uk.org/index.aspx

Crawford, R. H. and Treloar, G.J. (2004) Net energy analysis of solar and conventional domestic hot water systems in Melbourne, Australia. *Solar Energy,* **76**(1–3), 159–63.

CSIRO Manufacturing & Infrastructure Technology. *Embodied Energy.* http://www.cmit.csiro.au/brochures/tech/embodied/

Durrenberger, G., Patzel, N. and Hartmann, C. (2001) Household energy consumption in Switzerland. *International Journal of Environment and Pollution,* **15**(2), 159–70.

Glover, J., White, D. O. and Langrish, T. A. G. (2002) Wood versus concrete and steel in house construction: a life cycle assessment. *Journal of Forestry,* **100**(8), 34–41.

Hashimoto, S., Nose, M., Obara, T. and Moriguchi, Y. (2002) Wood products: potential carbon sequestration and impact on net carbon emissions of industrialized countries. *Environmental Science and Policy,* **5**, 183–93.

Jungbluth, N., Tieje, O. and Scholz, R. W. (2000) Food purchases: impacts from the consumers' point of view investigated with a modular LCA. *International Journal of Life Cycle Analysis,* **5**(3), 134–42.

Kunkel, K. E., Pielke Jr, R. A. and Changnon, S.A. (1999) Temporal fluctuations in weather and climate extremes that cause economic and human health impacts: a review. *Bulletin of the American Meteorological Society,* **80**(6), 1077–98. http://sciencepolicy.colorado.edu/admin/publication_files/resourse-75-1999.11.pdf

Natural Resources Canada (2003) *Energy Use Data Handbook 1990 and 1995 to 2001: Canada's natural resources now and in the future.* http://oee.nrcan.gc.ca/corporate/statistics/neud/dpa/data_e/Handbook04/Datahandbook2004.pdf

National Assessment Synthesis Team, US Global Change Research
     Program (2000) *Climate Change Impacts in the United States: The
     Potential Consequences of Climate Variability and Change*. Cam-
     bridge University Press, Cambridge.
Office of Energy Efficiency, Canada. *Statistics and Analysis*. http://
     oee.nrcan.gc.ca/corporate/statistics/neud/dpa/home.cfm?text=
     N&printview=N
Parry, M., Arnell, N., Hulme, M., Nicholls, R., and Livermore, M.
     (1998) Adapting to the inevitable. *Nature*, **395**, 741.
Reddy, B. V. V. and Jagadish, K. S. (2003) Embodied energy and alter-
     native building materials and technologies. *Energy and Buildings*,
     35(2), 129–37.
Rocky Mountain Institute. *Household Greenhouse Gas Emissions and
     Savings Measures*. http://www.rmi.org/sitepages/pid341.php
UK Energy Saving Trust. *My Home*. http://www.est.org.uk/myhome/
US Energy Information Administration. *Historical energy data for the
     US*. http://www.eia.doe.gov/neic/historic/hconsumption.htm
US Energy Information Administration. *Monthly energy review, US*.
     http://www.eia.doe.gov/emeu/mer/contents.html
US Energy Information Administration. *Residential energy consump-
     tion surveys, US*. http://www.eia.doe.gov/emeu/recs/contents.html
US Environmental Protection Agency and US Department of Energy.
     Energy Star. http://www.energystar.gov/
Wiel, S. and McMahon, J. E. (2003) Governments should implement
     energy-efficiency standards and labels – cautiously. *Energy Policy*,
     31, 1403–15.
Wilson, R. and Young, A. (1996) The embodied energy payback
     period of photovoltaic installations applied to buildings in the UK.
     Building and Environment, **31**(4), 299–305.

Chapter 4

Carlsson-Kanyama, A. (1998) Climate change and dietary choices –
     how can emissions of greenhouse gases from food consumption be
     reduced? *Food Policy*, **23**(3/4), 277–93.
Carlsson-Kanyama, A. *et al.* (2003) Food and life cycle energy inputs:
     consequences of diet and ways to increase efficiency. *Ecological Eco-
     nomics*, **44**, 293–307.

Hora, M. and Tick, J. (2001) *From Farm to Table: Making the Connection in the Mid-Atlantic Food System*. Capital Area Food Bank of Washington DC report.

Jones, A. (2001). *Eating Oil: Food Supply in a Changing Climate*. Sustain and Elm Farm Research Centre.

Jones, A. (2002) An environmental assessment of food supply chains: a case study of dessert apples. *Environmental Management*, **30**(4), 560–76.

Kramer, K. J. *et al.* (1999) Greenhouse gas emissions related to Dutch food consumption. *Energy Policy*, **27**, 203–16.

Lawrence, F. (2004) *Not on the Label: What Really Goes into the Food on Your Plate*. Penguin, London.

Parry, M., Rosenzweig, C., Iglesias, A., Fischer, G. and Livermore, M. (1999). *Global Environmental Change – Human and Policy Dimensions*, **9**: S51–S67 Supplement S.

Pirog, R., Van Plet, T., Enshayan, K. and Cook, E. (2001) Report for Leopold Center for Sustainable Agriculture, Iowa, US. http://www.leopold.iastate.edu/pubs/staff/papers.htm

Siikavirta, H. *et al.* (2003) Effects of e-commerce on greenhouse gas emissions: a case study of grocery home delivery in Finland. *Journal of Industrial Ecology*, **6**(2), 83–97.

Subak, S. (1999) Global environmental costs of beef production. *Ecological Economics*, **30**(1), 79–91.

Chapter 5

Australian Department of the Environment and Health (2001) *Independent assessment of kerbside recycling in Australia*, Volume 1. NOLAN-ITU Pty Ltd and Sinclair Knight Merz. Manly, NSW. http://www.deh.gov.au/industry/waste/covenant/kerbside.html

Bentham, G. (2002) Food poisoning and climate change. In Department of Health report – *Health Effects of Climate Change in the UK*, 4.2, pp. 81–98.

Bisgrove, R. and Hadley, P. (2002) *Gardening in the Global Greenhouse: The Impacts of Climate Change on Gardens in the UK*. Technical Report, UKCIP, Oxford.

Centre for Disease Control and Prevention. US Department of Health and Human Services. http://www.cdc.gov/

Environment Canada. *Canada's Greenhouse Gas Inventory 1990-2000*. Greenhouse Gas Division, Environment Canada. http://www.ec.gc.ca/pdb/ghg/1990_00_report/appa_e.cfm

Fehr, M. Cacado, M. D. R. and Romao, D. C. (2002) The basis of a policy for minimizing and recycling food waste. *Environmental Science and Policy*, **5**, 247–53.

Hayhoe, K. *et al.* (2004) Emissions pathways, climate change, and impacts on California. *Proceedings of the National Academy of Sciences of the United States of America*, **101**(34), 12422–7.

Hulme, M. (2003) Abrupt climate change: can society cope? *Philosophical Transactions of the Royal Society Series A*, **361**(1810), 2001–19.

NASA. Earth Observatory. http://earthobservatory.nasa.gov/

Parfitt, J. (2002) *Analysis of Household Waste Composition and the Factors Driving Waste Increases*. Strategy Unit, UK Government. http://www.number-10.gov.uk/su/waste/report/downloads/composition.pdf

Parliamentary Office of Science and Technology (2004) UK health impacts of climate change. *POSTnote* Number 232.

Pickin, J. G., Yuen, S. T. S. and Hennings, H. (2002) Waste management options to reduce greenhouse gas emissions from paper in Australia. *Atmospheric Environment*, **36**(4), 741–52.

Reay, D. S. (2003) Sinking methane. *Biologist*, **50**(1), 15–19.

UK Department of Trade and Industry (2002) *Environmental Life Cycle Assessment and Financial Life Cycle Analysis of the WEEE Directive and its Implications for the UK*. Report prepared by PriceWaterhouseCoopers. http://www.dti.gov.uk/support/dtiweeeupdate.pdf

US Energy Information Administration (2003) *Emissions of Greenhouse Gases in the United States 2003: Methane Emissions*. http://www.eia.doe.gov/oiaf/1605/ggrpt/methane.html

US Environmental Protection Agency. *Greenhouse Gas Emissions from Management of Selected Materials in Municipal Solid Waste*. Washington, DC. http://www.epa.gov/epaoswer/non-hw/muncpl/ghg/chapter4.pdf

US Environmental Protection Agency. *WasteWise: Changing with Climate*. Washington, DC.

US Environmental Protection Agency (2003) *Municipal Solid Waste in the United States: 2001 Facts and Figures*. Office of Solid Waste and Emergency Response. Washington, DC.

WasteWatch. *WasteOnline: In Depth Information on Waste*. http://www.wasteonline.org.uk/

Weitz, K. A., Thorneloe, S. E., Nishtala, S. R., Yarosky, S. and Zannes, M. (2002) The impact of municipal solid waste management on greenhouse gas emission in the United States. *Journal of the Air and Waste Management Association*, **52**(9), 1000–11.

Williams, I. D. and Kelly, J. (2003) Green waste collection and the public's recycling behaviour in the Borough of Wyre, England. *Resources, Conservation and Recycling*, **38**, 139–59.

Chapter 6

Adams, D. and Carwardine, M. (1991) *Last Chance to See*. Pan Macmillan, London.

Allen, M.R. (2004) The Blame Game: Who will pay for the damaging consequences of climate change? *Nature* **432**, 551–2

Anderson, K. and Starkey, R. (2004) *Domestic Tradable Quotas: a policy instrument for the reduction of greenhouse gas emissions*. An Interim Report to the Tyndall Centre for Climate Change Research. Tyndall North, Manchester.

Australian Greenhouse Office (2002) *Living With Climate Change – An Overview of Potential Climate Change Impacts on Australia*. http://www.greenhouse.gov.au/impacts/overview/

Barker, T. and Ekins, P. (2004) The costs of Kyoto for the US economy. *Energy Journal*, **25**(3), 53–71.

Broadmeadow, M. (2000) Climate change – implications for forestry in Britain. *Forestry Commission Bulletin*, 125. Forestry Commission UK.

Burke, L. and Maidens, J. (2004) *Reefs at Risk in the Caribbean*. World Resources Institute. http://pdf.wri.org/reefs_caribbean_full.pdf

Clarkson, R. and Deyes, K. (2002) *Estimating the social cost of carbon emissions*. Government Economic Service Working Paper 140. http://www.hm-treasury.gov.uk/media/209/60/SCC.pdf

DeLeo, G. A., Rizzi, L., Caizzi, A. and Gatto, M. (2001) The economic benefits of the Kyoto Protocol. *Nature*, **413**, 478–9.

Dresner S. and Ekins, P. (2004) *Economic Instruments for a Socially Neutral National Home Energy Efficiency Programme*. Policy Studies Institute Research Discussion Paper 18. http://www.psi.org.uk/docs/rdp/rdp18-dresner-ekins-energy.pdf

Dresner, S. and Ekins, P. (2004) *The Distribution Impacts of Economic Instruments to Limit Greenhouse Gas Emissions from Transport.* Policy Studies Institute Research Discussion Paper 19. http://www.psi.org.uk/docs/rdp/rdp19-dresner-ekins-transport.pdf

Dresner S. and Ekins, P. (2004) *Charging for Domestic Waste: Combining Environment and Equity Considerations.* Policy Studies Institute Research Discussion Paper 20. http://www.psi.org.uk/docs/rdp/rdp20-dresner-ekins-waste.pdf

Insure.com (2002) *10 Years Later, Hurricane Andrew Would Cost Twice as Much.* http://info.insure.com/home/disaster/andrewtoday/

Howarth, R. B. (2001) Intertemporal social choice and climate stabilization. *International Journal of Environment and Pollution*, **15**(4), 386–405.

Kamal, W. A. (1997) Improving energy efficiency – the cost-effective way to mitigate global warming. *Energy Conversion and Management*, **38**(1), 39–59.

Ogden, J. M., Williams, R. H. and Larson, E. D. (2004) Societal lifecycle costs of cars with alternative fuels/engines. *Energy Policy*, **32**(1), 7–27.

O'Hara, M. (2004) Homeowners face a rising tide. Jobs and Money, *The Guardian*, 14 February, p. 9.

Stott, P. A., Stone, D. A. and Allen, M. R. (2004) Human contribution to the European heatwave of 2003. *Nature*, **432**, 610–14.

Thomas, C. D. *et al.* (2004) Extinction risk from climate change. *Nature*, **427**, 145–8.

Tol, R. S. J. and Verheyen, R. (2004) State responsibility and compensation for climate change damages – a legal and economic assessment. *Energy Policy*, **32**(9), 1109–30.

US Energy Information Administration (1998) *Impacts of the Kyoto Protocol on US Energy Markets and Economic Activity.* US Department of Energy, Washington, DC. http://www.eia.doe.gov/oiaf/kyoto/pdf/sroiaf9803.pdf

US Environmental Protection Agency (2003) *Pay-As-You-Throw: A Cooling Effect On Climate Change.* http://www.epa.gov/mswclimate/

Yohe, G., Neumann, J., Marshall, P. and Ameden, H. (1996) The economic cost of greenhouse-induced sea-level rise for developed property in the United States. *Climatic Change*, **32**(4), 387–410.

Chapter 7

Clark, T. *Greening Your Final Arrangements*. Jewish-Funerals.org http:/ /www.jewish-funerals.org/greeningfinal.htm

Clean Air – Cool Planet. http://www.cleanair-coolplanet.org/

Linderhof, V. G. M. (2001) Household demand for energy, water and the collection of waste: a microeconometric analysis. *PhD Thesis*, Rijksuniversiteit, Groningen. Labyprint Publication, Holland.

Lyman, F. (2003) Green graves give back to nature: eco-friendly funerals break new ground. *MSNBC News*. http://msnbc.msn.com/ id/3076642/

Rosen, K. B. and Meier, A. K. (1999) *Energy use of televisions and video-cassette recorders in the US*. US Department of Energy. http:// eetd.lbl.gov/EA/Reports/42393/42393.pdf

US Energy Information Administration. *Historical End-Use Consumption Data*. http://www.eia.doe.gov/neic/historic/hconsumption.htm

US Energy Information Administration. *Environment: Energy Related Emissions Data, Forecasts and Analyses*. http://www.eia.doe.gov/ environment.html

US Energy Information Administration (2005) *Annual Energy Outlook 2005 with Projections to 2025*. Report #: DOE/EIA-0383(2005). http://www.eia.doe.gov/oiaf/aeo/

Worrell, E., Price, L., Martin, N., Hendriks, C. and Meida, L. O. (2001) Carbon dioxide emissions from the global cement industry. *Annual Review of Energy and the Environment*, **26**, 303–29.

Chapter 8

Australian Greenhouse Office (2005) *Buildings and Energy: Office Building Energy Use*. http://www.greenhouse.gov.au/lgmodules/ wep/buildings/training/training4.html

Australian Greenhouse Office. *National Energy Star*. http:// www.energystar.gov.au/

Climate Ark. http://www.climateark.org/

Community for Environmental Engineering and Technology in Australia. http://www.comeeta.org/

Envirowise and the UK Environment Agency. *Green Officiency: Running a Cost-Effective, Environmentally Aware Office*. GG256. Envirowise, Oxfordshire.

National Appliance and Equipment Energy Efficiency Committee (NAEEEC). *Green Office Guide: A Guide to Help You Buy and Use Environmentally Friendly Office Equipment.* http://www.energystar.gov. au/consumers/greenbook.html

Picklum, R. E., Nordman, B. and Kresch, B. (1999) *Guide to Reducing Energy Use in Office Equipment.* US Department of Energy. http:// eetd.lbl.gov/bea/sf/GuideR.pdf

Sellen, A. J. and Harper, R. H. R. (2001) *The Myth of the Paperless Office.* MIT Press, Cambridge MA.

The Guardian (2004). *Green Offices.* http://www.guardian.co.uk/ values/socialaudit/environment/story/0,15074,1305103,00.html

UK Department of the Environment and Transport and the Regions (DETR) (2000) *Climate Change: the UK Program.* http://www.defra. gov.uk/environment/climatechange/cm4913/index.htm#docs

US Energy Information Agency. *Information on the commercial buildings sector.* http://www.eia.doe.gov/emeu/cbecs/contents.html

US Environmental Protection Agency. *Planning and Urban Environment.* http://yosemite.epa.gov/OAR/globalwarming.nsf/content/ ActionsLocalSmartSavingsPlanningandUrbanEnvironment.html

World Resources Program (2001) No end to paperwork. Vanasselt, W. (ed.) *Earthtrends*, World Resources Institute. http://earthtrends.wri. org/pdf_library/features/ene_fea_paper.pdf

bibliography

Adams, D. and Carwardine, M. (1991) *Last Chance to See*. Pan Macmillan, London.

Bunting, M. (2004) *Willing Slaves: How the Overwork Culture is Ruling Our Lives*. HarperCollins, London.

Dauncey, G. and Mazza, P. (2001) *Stormy Weather: 101 Solutions to Global Climate Change*. New Society Publishers, Canada.

Diamond, J. (2005) *Collapse: How Societies Choose to Fail or Survive*. Allen Lane-Penguin, London.

Hilman, M. and Fawcett, T. (2004) *How We Can Save the Planet*. Penguin, London.

Houghton, J. (1997) *Global Warming: The Complete Briefing*. Cambridge University Press, Cambridge.

Intergovernmental Panel on Climate Change (IPCC). (2000) *Aviation and the Global Atmosphere*. Cambridge University Press, Cambridge.

Intergovernmental Panel on Climate Change (IPCC) (2001) *Climate Change 2001: The Scientific Basis*. Cambridge University Press, Cambridge.

Intergovernmental Panel on Climate Change (IPCC) (2001) *Climate Change 2001: Impacts, Adaptation, and Vulnerability*. Cambridge University Press, Cambridge.

Intergovernmental Panel on Climate Change (IPCC) (2001) *Climate Change 2001: Mitigation*. Cambridge University Press, Cambridge.

Jones, A. (2001) *Eating Oil: Food Supply in a Changing Climate*. Sustain and Elm Farm Research Centre, London.

Langholz, J. and Turner, K. (2003) *You Can Prevent Global Warming (and Save Money)*. Andrews McMeel Publishing, Kansas City.

Lawrence, F. (2004) *Not on the Label: What Really Goes into the Food on Your Plate*. Penguin, London.

Lynas, M. (2004) *High Tide: News from a Warming World*. Flamingo, London.

Meyer, A. (2000) *Contraction & Convergence: The Global Solution to Climate Change*. Schumacher Briefings, Green Books, Devon.

Monbiot, G. (2004) *The Age of Consent: A Manifesto for a New World Order*. HarperCollins, London.

National Assessment Synthesis Team, US Global Change Research Program (2000) *Climate Change Impacts in the United States: The Potential Consequences of Climate Variability and Change*. Cambridge University Press, Cambridge.

Smith, A. and Baird, N. (2005) *Save Cash & Save the Planet*. HarperCollins, London.

index

THE WHOLE STORY

ALTERNATIVE MEDICINE ON TRIAL?

TOBY MURCOTT

THE WHOLE STORY
ALTERNATIVE MEDICINE ON TRIAL?
by Toby Murcott
MACMILLAN; ISBN: 1–4039–4500–4; £16.99/$24.95; HARDCOVER

"Essential reading. A balanced, sympathetic and long overdue look at the relationship between science and complementary medicine." *Focus Magazine*

"This book should be prescribed for bigots on both sides, to be taken thoughtfully, all the way to the last page." *Guardian*

order now from www.macmillanscience.com

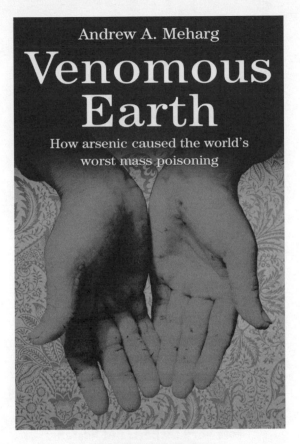

VENOMOUS EARTH
HOW ARSENIC CAUSED THE WORLD'S WORST MASS
POISONING
by Andrew Meharg
MACMILLAN; ISBN: 1–4039–4499–7 £16.99/US$29.95;
HARDCOVER

"Meharg is good on the technological and political challenges of testing water. He is terrific on the wider history of arsenic, in alchemy, industry and interior decorating." *Guardian*

"Meharg tells the lively and cautionary story of arsenic's misuse over the centuries." *Newsweek International*

order now from www.macmillanscience.com

MAX BORN

THE
BORN–EINSTEIN
LETTERS
1916–1955

Friendship, Politics and Physics
in Uncertain Times

Introduction by **Werner Heisenberg** Foreword by **Bertrand Russell**
New Preface by **Diana Buchwald and Kip Thorne**

THE BORN–EINSTEIN LETTERS
FRIENDSHIP, POLITICS AND PHYSICS IN UNCERTAIN TIMES
by Max Born and Albert Einstein
Introduction by Werner Heisenberg
Foreword by Bertrand Russell
New Preface by Diana Buchwald and Kip Thorne
MACMILLAN; ISBN: 1–4039–4496–2; £19.99/US$26.95;
HARDCOVER

"An immensely readable personal account of Einstein's struggles with other physicists." *Washington Post*

"With a well-informed introductory essay by Buchwald and Thorne, the correspondence is a delight, enabling us to trace the development of the intriguing friendship between the two physicists and to read their views on the great themes of physics and politics of their time." *The Times Higher Education Supplement*

order now from www.macmillanscience.com

Mathematics with Love

The Courtship Correspondence of
Barnes Wallis,
Inventor of the Bouncing Bomb

Mary Stopes-Roe

MATHEMATICS WITH LOVE
THE COURTSHIP CORRESPONDENCE OF BARNES WALLIS,
INVENTOR OF THE BOUNCING BOMB
by Mary Stopes-Roe
MACMILLAN; ISBN 1–4039–4498–9; £19.99/US$24.95;
HARDCOVER

"In place of poetry and roses, engineer Barnes Wallis wooed his lady-love with trigonometry and calculus – and won her heart. A charming and unique correspondence from the human side of mathematics." *Ian Stewart, author of* Math Hysteria *and* Flatterland

"One of the tenderest and most unlikely courtships imaginable... A heartwarming insight into the selfless and truly romantic way that relationships were forged between the wars." *Daily Mail*

order now from www.macmillanscience.com

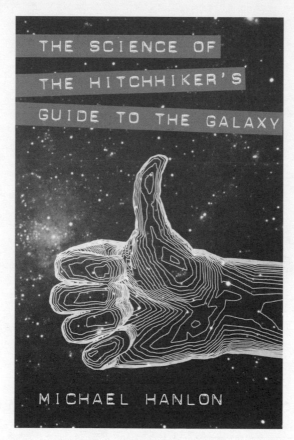